Thomas Box

Practical hydraulics: a series of rules and tables for the use of engineers

Second Edition

Thomas Box

Practical hydraulics: a series of rules and tables for the use of engineers
Second Edition

ISBN/EAN: 9783337157401

Printed in Europe, USA, Canada, Australia, Japan

Cover: Foto ©berggeist007 / pixelio.de

More available books at **www.hansebooks.com**

PRACTICAL HYDRAULICS:

A SERIES

OF

RULES AND TABLES

FOR

THE USE OF ENGINEERS, Etc., Etc.

BY

THOMAS BOX,

Author of 'Practical Treatise on Heat.' 'Mill-gearing,' etc.

SECOND EDITION.

LONDON:
E. & F. N. SPON, 48, CHARING CROSS.
1870.

PREFACE TO THE SECOND EDITION.

In preparing a Second Edition of 'Practical Hydraulics' considerable alterations and additions have been made. To facilitate reference, the work has been divided into Chapters; additional Rules for Culverts and other subjects have been given, including several new Tables, and an increased number of Illustrations. These alterations were so considerable, that it was found necessary to re-write the whole, and thus opportunity was given to introduce much new and valuable information, which, it is hoped, will increase the usefulness of the work.

Bath, *July*, 1870.

PREFACE TO THE FIRST EDITION.

The reader must not expect, in this little book, an exhaustive treatise on Hydraulics; many such have been written, and they leave little or nothing to be desired. This work consists of a series of Rules and Tables, giving unusual facility for the solution of questions which occur in the daily practice of Engineers.

For the two leading questions—the Discharge of Pipes, and of Open Channels—two sets of Tables are given, the reason for which may not be obvious; but it is impossible to give Tables combining extreme facility with extreme accuracy for low heads, and the author has therefore given two Tables, one giving accurate results in all ordinary cases with the least possible labour, and the other giving, with more labour, exact results in extreme cases.

For the most part the Rules and Tables have been long used in an extensive practice, and the principal reason for publishing them is the author's desire that the profession from which he has retired may have the benefit of Tables, &c., which for many years have been very useful to himself.

EASEDALE, GRASMERE,
July, 1867.

CONTENTS.

CHAPTER I.—On the Discharge of Apertures, Pipes, &c.

	PAGE
Velocity of Efflux	1
Discharge by an Orifice in a Thin Plate	1
,, by Short Tubes	3
Friction of Long Pipes	4
Head for Velocity of Entry	18
Bends, loss of Head by	19
Compound Water-mains, Discharge of	22
Effect of Contour of Section	24
Special Cases, Examples of	27
Delivery and Suction-pipes to Pumps	28
Service-pipes in Towns	29
General Laws for Pipes	30
Head for very Low Velocities, by Prony's Formula	31
Square and Rectangular Pipes	34
Effect of Corrosion or Rust in Pipes	35

CHAPTER II.—On Fountains, Jets, &c.

Height of Jets with given Heads	37
Discharge of Jets	40
Jets at the End of Long Mains	40
Path of Fountain Jets	43
Ornamental Jets	47

CHAPTER III.—On Canals, Culverts, and Water-courses.

Open Water-courses	48
Head due to Velocity in Open Channels	49
,, to overcome Friction in Long Channels	50
River Channels of irregular Cross-section	54
Openings of Bridges	55

CONTENTS.

	PAGE
Submerged Openings	55
Discharge of Oval Culverts	57
Head for very Low Velocities—Eytelwein's Formula	57
Case of a Mill-stream	61

CHAPTER IV.—ON WEIRS, OVERFLOW-PIPES, &c.

Weirs, Discharge and Form of	63
,, Effect of Thickness of Crest on the Discharge	66
,, ,, of Velocity of Approach on ,,	67
Correction for Short Weirs	67
Overflow-pipes to Tanks	69
,, to Fountains	71
,, Common	71

CHAPTER V.—ON THE STRENGTH AND PROPORTIONS OF WATER-PIPES—RAINFALL, &c., &c.

Strength of Thick Pipes by Barlow's Rule, &c.	72
,, Thin ,,	73
Proportions and Weights of Cast-iron Socket-pipes	75
,, of Flange-pipes	76
Strength of Lead Pipes	78
Power of Horses, &c., in raising Water	78
Rainfall—Heavy Rains—Rain-water Tanks	79
Weight and Pressure of Water	80

PRACTICAL HYDRAULICS.

CHAPTER I.

DISCHARGE OF APERTURES, PIPES, &C.

(1.) "*Velocity of Efflux.*"—The velocity with which water issues from the side of a vessel, as at A, Fig. 1, is the same as that of a body falling freely by gravity from the height H, or the distance from the centre of the orifice to the surface of the water. This velocity is given by the rule:—

$$V = \sqrt{H} \times 8$$

In which H = the height or head of water in feet, and V = the velocity in feet per second. From this we may obtain another rule giving the discharge in gallons, which becomes:—

$$G = \sqrt{H} \times d^2 \times 16\cdot 3$$

In which H = the head of water in feet, d = the diameter of the orifice in inches, and G = gallons discharged per minute. Table 1 has been calculated by this rule.

These rules give the *theoretical* velocity and discharge; for application to practice, they may require some modification to adapt them to the particular form of the orifice.

(2.) "*Discharge by an Orifice in a Thin Plate.*"—It has been found by experiment that, when the discharging orifice is made in a thin plate, the converging currents of water approaching the aperture cause a *contraction* in the issuing stream, so that instead of a parallel or cylindrical jet, it becomes a conical one of the form shown by Fig. 2, the greatest contraction being at

THEORETICAL DISCHARGE OF APERTURES.

TABLE 1.—Of the Theoretical Discharge of Water by Round Apertures of various Diameters, and under Different Heads of Water Pressure.

Head of Water in Inches.

Discharge in Gallons per Minute.

Diam. in Inches.	1	2	3	4	5	6	7	8	9	10	12	14	16	18	20	22	24
1	4·7	6·6	8·1	9·4	10·5	11·5	12·4	13·3	14·1	14·8	16·2	17·6	18·8	19·9	21	22	23
2	18·8	26·4	32·4	37·6	42·0	46·0	49·6	53·2	56·4	59·2	64·4	70·4	75·2	79·6	84	88	92
3	42·2	59·4	72·9	84·6	94·5	103	112	120	127	133	146	158	169	179	189	198	207
4	75·2	106	130	150	168	184	198	213	225	237	259	281	301	318	336	352	368
5	117	165	203	235	262	287	310	332	352	370	405	440	470	497	525	550	575
6	169	237	291	338	378	414	446	479	507	533	583	663	677	716	756	792	828
7	230	310	397	460	514	563	607	652	691	725	794	862	921	975	1029	1078	1127
8	301	422	518	601	672	736	793	851	902	947	1037	1126	1203	1273	1344	1408	1472
9	381	534	656	761	850	931	1006	1077	1142	1199	1312	1425	1523	1612	1701	1782	1863
10	470	660	810	940	1050	1150	1240	1330	1411	1480	1620	1760	1880	1990	2100	2200	2300
12	676	952	1168	1353	1512	1656	1785	1915	2030	2134	2333	2534	2707	2865	3024	3170	3312
14	920	1241	1588	1842	2058	2254	2430	2606	2764	2901	3175	3450	3684	3900	4116	4312	4508
16	1203	1690	2074	2406	2688	2944	3174	3405	3610	3789	4147	4506	4813	5094	5376	5632	5888
18	1523	2138	2624	3045	3402	3726	4018	4309	4568	4795	5249	5702	6091	6447	6804	7128	7452
20	1880	2640	3240	3760	4200	4600	4960	5320	5640	5920	6480	7040	7520	7960	8400	8800	9200
22	2275	3194	3920	4550	5082	5566	6002	6437	6824	7163	7841	8518	9099	9632	10164	10648	11132
24	2704	3808	4672	5414	6048	6624	7140	7660	8120	8536	9332	10136	10829	11460	12096	12680	13248
30	4230	5940	7290	8460	9450	10350	11160	11970	12630	13320	14580	15840	16320	17910	18900	19800	20700
Velocity in feet per second	2·32	3·275	4·01	4·63	5·18	5·67	6·13	6·55	6·95	7·32	8·03	8·67	9·27	9·83	10·36	10·87	11·35

the point C, whose distance from the plate is half the diameter of the orifice, and its diameter ·784, that of the orifice being 1. The form from B to C may be taken as a curve, whose radius is 1·22 times the diameter of the orifice.

Now, the foregoing rule gives the maximum velocity, or that at the point of greatest contraction C, and if the diameter be taken there, the rules would give the true velocity and discharge without correction. But it is obvious that the velocity at the aperture itself (or at B) would be less than at C in the ratio of the respective areas at the two points, or as 1^2 to $·784^2$ or 1 to ·615, and in that case, the diameter being taken at B, the velocity there would become $V = \sqrt{H} \times 8 \times ·615$ and the discharge $G = \sqrt{H} \times d^2 \times 16·3 \times ·615$. From this we get for apertures in a thin plate, the rules:—

$$G = \sqrt{H} \times d^2 \times 10$$

$$H = \left(\frac{G}{d^2 \times 10}\right)^2$$

$$d = \left(\frac{G}{\sqrt{H} \times 10}\right)^{\frac{1}{2}}$$

Thus, with 3 inches diameter and 16 feet head, the discharge would be $\sqrt{16} \times 3^2 \times 10$, or $4 \times 9 \times 10 = 360$ gallons per minute. The head for 150 gallons per minute with 2 inches diameter $= \left(\frac{150}{4 \times 10}\right)^2 = 14·06$ feet; and the diameter for 200 gallons per minute with 20 feet head would be $\left(\frac{200}{4·47 \times 10}\right)^{\frac{1}{2}} = 2·11$ inches, &c., &c.

(3.) "*Discharge by Short Tubes.*"—When the aperture is of considerable thickness, or has the form of a short tube, not less in length than twice the diameter, the amount of contraction is found to be less, and the discharge greater, than with a thin plate. Fig. 3 shows a tube 1 inch diameter and 2 inches long; the greatest contraction is in that case ·9 inch diameter, and its proportional area $·9^2 = ·81$, or say $·8$ of the area of the tube. For short tubes therefore the rules become:—

$$G = \sqrt{H} \times d^2 \times 13$$
$$H = \left(\frac{G}{d^2 \times 13}\right)^2.$$
$$d = \left(\frac{G}{\sqrt{H} \times 13}\right)^{\frac{1}{2}}$$

Table 2 has been calculated by these rules; thus, for a 7-inch pipe discharging 450 gallons, the Table shows that the head necessary to generate the velocity at entry is 6 inches; this is irrespective of friction, which, in fact, for so short a tube as the rule supposes, would be practically nothing. This Table applies to all cases of pipes; for instance, Fig. 4 shows the inlet end of a main from a reservoir, which will require for the velocity at entry alone the amount of head shown by the Table. When, as is usually the case, the pipe is of considerable length, the head due to friction must also be allowed for.

(4.) "*Friction of Long Pipes.*"—With a long pipe there is not only the loss of head due to the velocity at entry, but also another loss due simply to the friction of the water against the sides of the pipe, so that in all cases the head consumed may be considered as composed of two portions:—one, the amount due to velocity of entry, irrespective of friction; and the other, the amount due to friction alone. Thus, in Fig. 8 the head h gives a certain velocity of discharge by the short pipe A; but to give the same velocity in the long main B C, the head H' is necessary, of which h' is consumed in generating the velocity at entry, being the same as for A, and the rest, or H, in the friction of the long pipe: the total head is, of course, the sum of the two.

(5.) The loss of head by friction may be calculated by the following rules:—

$$G = \left(\frac{(3d)^5 \times H}{L}\right)^{\frac{1}{2}}$$
$$H = \frac{G^2 \times L}{(3d)^5}$$

TABLE 2.—Of the Actual Discharge by Short Tubes of various Diameters, with Square Edges and under Different Heads of Water Pressure, being 10/13ths of the Theoretical Discharge.

DISCHARGE BY SHORT TUBES.

Diam. in Inches.	Head of Water in Inches. Discharge in Gallons per Minute.																
	1	2	3	4	5	6	7	8	9	10	12	14	16	18	20	22	24
1	3·76	5·28	6·48	7·52	8·4	9·2	9·9	10·6	11·3	11·8	13·0	14·1	15·0	15·9	16·8	17·6	18·4
2	15·04	21·12	25·9	30·1	33·6	36·8	39·7	42·6	45·1	47·4	51·8	56·3	60·2	63·7	67·2	70·4	73·6
3	33·8	47·5	58·3	67·7	75·6	82·4	89·6	96·0	101·6	106·4	116·8	126	135	143	151	158	166
4	60·2	84·8	104	120	130	147	158	170	180	189	207	225	241	254	269	282	294
5	93·6	132	162	188	210	230	248	266	282	296	324	352	376	398	420	440	460
6	135	190	233	270	302	331	357	382	406	426	466	530	542	573	605	634	662
7	194	248	318	368	411	450	486	522	553	580	636	689	737	780	823	862	902
8	241	338	414	481	538	589	634	681	722	758	829	901	962	1018	1075	1126	1178
9	305	427	525	609	680	745	805	863	914	959	1049	1140	1218	1290	1361	1426	1490
10	376	528	648	732	840	920	992	1064	1129	1184	1296	1408	1504	1592	1680	1760	1840
12	541	762	934	1082	1210	1325	1428	1532	1624	1707	1866	2027	2166	2292	2419	2536	2650
14	736	993	1268	1474	1646	1803	1944	2085	2211	2321	2540	2760	2947	3120	3293	3450	3606
15	846	1188	1458	1692	1890	2070	2232	2394	2288	2664	2916	3168	3384	3582	3780	3960	4140
16	962	1352	1659	1925	2150	2355	2539	2724	2888	3031	3318	3605	3850	4075	4301	4406	4710
18	1218	1710	2099	2436	2722	2981	3214	3447	3662	3836	4199	4562	4873	5158	5443	5702	5962
20	1504	2112	2592	3008	3360	3680	3968	4256	4512	4736	5184	5632	6016	6368	6720	7040	7360
22	1820	2552	3136	3640	4065	4452	4801	5149	5459	5730	6272	6814	7279	7705	8131	8518	8905
24	2163	3046	3737	4331	4838	5299	5712	6128	6496	6828	7465	8108	8663	9168	9676	10144	10598
30	3384	4752	5832	6768	7560	8280	8928	9576	10152	10656	11664	12672	13536	14328	15120	15840	16560

$$d = \left(\frac{G^2 \times L}{H}\right)^{\frac{1}{5}} \div 3$$

$$L = \frac{(3d)^5 \times H}{G^2}$$

In these rules d = diameter of the pipe in inches.
$\quad\quad\quad\quad\;\; L$ = length in yards.
$\quad\quad\quad\quad\;\; H$ = head of water in feet.
$\quad\quad\quad\quad\;\; G$ = gallons per minute.

These rules require the use of logarithms to work them easily: thus, to find the discharge by a 7-inch pipe 3797 yards long with 45 feet head, we have:—

$$
\begin{array}{r}
7 \times 3 = 21 = 1\cdot322219 \\
5 \\
\hline
6\cdot611095 \\
\times\; 45 = 1\cdot653213 \\
\hline
8\cdot264308 \\
\div\; 3797 = 3\cdot579441 \\
\hline
2)4\cdot684867 \\
\hline
2\cdot342433 = 220 \text{ gallons per minute.}
\end{array}
$$

Again, to find the head necessary to discharge 320 gallons per minute by an 8-inch pipe 3457 yards long, we have:—

$$
\begin{array}{r}
320 = 2\cdot505150 \\
2 \\
\hline
5\cdot010300 \\
\times\; 3457 = 3\cdot538699 \\
\hline
8\cdot548999 \\
8 \times 3 = 24 = 1\cdot380211 \times 5 = 6\cdot901055 \\
\hline
1\cdot647944 = 44\cdot46 \text{ feet head.}
\end{array}
$$

And again, to find the diameter for 110 gallons per minute with 56 feet head, the length being 273 yards, we have:—

$$110 = 2\cdot041393$$
$$\underline{2}$$
$$4\cdot082786$$
$$\times\ 273 = 2\cdot436163$$
$$\overline{6\cdot518949}$$
$$\div\ 56 = 1\cdot748188$$
$$\overline{5)4\cdot770761}$$

$\cdot954152 = 9$, and $\dfrac{9}{3} = 3$ inches diameter.

Table 3 has been calculated by these rules, and will greatly facilitate the calculation of pipe questions, it also has the great advantage of requiring only the simple rules of arithmetic.

(6.) 1st. Having G, L, and d given, to find H. In the Table opposite the given number of gallons, and under the given diameter, is found the head due to a length of one yard, and multiplying that number by the given length in yards, gives the required head of water in feet. Thus, taking our former illustration in (5), the head to deliver 320 gallons per minute by an 8-inch pipe 3457 yards long—opposite 320 gallons in the Table, and under 8 inches diameter, is ·01286 feet, and ·01286 × 3457 = 44·46 feet, the head sought.

(7.) 2nd. To find d, having H, L, and G given. Divide the given head of water in feet by the given length in yards, and the nearest number thereto in the Table opposite the given number of gallons will be found under the required diameter. Thus, to find, the diameter for 110 gallons per minute with 56 feet head, the length being 273 yards, we have $\dfrac{56}{273} = \cdot 205$, looking for which in the Table opposite 110 gallons we find it under 3 inches, the diameter sought (see 5). Again, to find the diameter for 320 gallons, 20 feet head, and 1600 yards long, we have $\dfrac{20}{1600} = \cdot 0125$, the nearest number to which, in the Table (·01286) is found under 8 inches, the diameter sought. In most cases the tabular number will not be the exact number

TABLE 3.—Of the Head of Water consumed by Friction with Pipes 1 yard long.

Gallons per Minute	Diameter of the Pipe in Inches						
	1	1½	2	2½	3	3½	4
	Head of Water in Feet.						
1	·0041	·00054	·00012	·000042	·000016	·0000078	·000004
2	·0164	·00216	·00051	·000168	·000067	·0000313	·000016
3	·0370	·00487	·00115	·000379	·000152	·0000705	·000036
4	·0658	·00867	·00205	·000674	·000271	·000125	·000064
5	·1028	·01354	·00321	·001053	·000423	·000195	·000100
6	·1481	·01950	·00463	·001517	·000609	·000282	·000144
7	·2016	·02655	·00630	·002064	·000830	·000383	·000196
8	·2633	·03468	·00823	·002696	·001084	·000501	·000257
9	·3333	·04389	·01041	·003413	·001372	·000634	·000325
10	·411	·0541	·01286	·00421	·00169	·000783	·000401
20	1·64	·2167	·0514	·01685	·00577	·00313	·00160
30	3·70	·4877	·115	·03792	·0152	·00707	·00361
40	6·58	·8670	·205	·06742	·0271	·01253	·00643
50	10·28	1·35	·321	·1053	·0423	·01958	·01004
60	14·81	1·95	·463	·1517	·0609	·02820	·01446
70	20·16	2·65	·630	·2064	·0830	·03839	·01969
80	26·33	3·46	·823	·2696	·1084	·05014	·02572
90	33·33	4·38	1·041	·3413	·1372	·06346	·03255

Gallons per Minute	Diameter of the Pipe in Inches.						
	5	6	7	8	9	10	12
	Head of Water in Feet.						
10	·000131	·000052	·000024	·000012	·0000069	·00000411	·00000165
20	·000526	·000211	·000097	·000050	·0000278	·00001646	·00000661
30	·001185	·000476	·000220	·000113	·0000627	·00003703	·00001488
40	·002003	·000804	·000372	·000191	·0001060	·00006259	·00002515
50	·003292	·001323	·000612	·000314	·0001742	·0001028	·0000413
60	·004741	·001905	·000881	·000452	·0002569	·0001481	·0000595
70	·006453	·002593	·001200	·000616	·0003415	·0002016	·0000810
80	·008428	·003386	·001567	·000803	·0004460	·0002533	·0001058
90	·010637	·004286	·001983	·001017	·0005645	·0003333	·0001339

Note.—For intermediate numbers, see body of general Table 3, as explained in (10) page 16.

HYDRAULIC TABLE 3—continued.

HEAD FOR FRICTION OF LONG PIPES.

Gallons per Minute.	\multicolumn{9}{c}{Diameter of the Pipe in Inches.}								
	1	1¼	2	2½	3	3½	4	5	6	7	8	9
						Head of Water in Feet.						
100	41·1	5·4	1·28	·421	·169	·078	·0401	·01317	·005292	·00244	·001256	·00069
110	49·7	6·5	1·55	·509	·205	·094	·0486	·01539	·006403	·00296	·001519	·00084
120	59·2	7·8	1·85	·606	·243	·112	·0578	·01896	·007620	·00352	·001808	·00100
130	69·5	9·1	2·17	·712	·286	·132	·0679	·02225	·008443	·00413	·002122	·00117
140	80·6	10·6	2·52	·825	·332	·153	·0788	·02581	·010372	·00480	·002461	·00136
150	92·5	12·1	2·89	·948	·381	·176	·0904	·02963	·011907	·00551	·002826	·00156
160	105·3	13·8	3·29	1·078	·433	·200	·1028	·03371	·013547	·00626	·003215	·00178
170	118·9	15·6	3·71	1·217	·485	·226	·1161	·03806	·015293	·00707	·003629	·00201
180	133·3	17·5	4·16	1·365	·549	·253	·1312	·04267	·017146	·00793	·004069	·00225
190	148·5	19·5	4·64	1·521	·611	·282	·1450	·04754	·019104	·00884	·004534	·00251
200	164·6	21·6	5·14	1·685	·677	·313	·1607	·05268	·021168	·00979	·005024	·00278
210	181·4	23·8	5·67	1·858	·747	·345	·1772	·05807	·023337	·01080	·005538	·00307
220	199·1	26·2	6·22	2·039	·819	·379	·1945	·06374	·025613	·01185	·006079	·00337
230	217·6	28·6	6·80	2·229	·896	·414	·2126	·06966	·027995	·01295	·006644	·00368
240	237·0	31·2	7·40	2·427	·975	·451	·2314	·07585	·030482	·01410	·007234	·00401
250	257·1	33·8	8·03	2·633	1·058	·489	·2511	·08231	·033075	·01530	·007850	·00435
260	278·1	36·6	8·69	2·848	1·145	·529	·2716	·08902	·035773	·01655	·008490	·00471
270	299·9	39·5	9·37	3·071	1·234	·571	·2929	·09600	·038578	·01785	·009156	·00508
280	322·6	42·4	10·08	3·303	1·328	·614	·3150	·10325	·041489	·01920	·009847	·00546
290	346·0	45·5	10·81	3·544	1·424	·658	·3379	·11075	·044506	·02039	·010562	·00586
300	370·3	48·7	11·58	3·792	1·524	·705	·3617	·11853	·047628	·02204	·011304	·00627
310	395·4	52·0	12·35	4·049	1·627	·752	·3862	·12655	·050856	·02353	·012070	·00669

HYDRAULIC TABLE 3—continued.

HEAD FOR FRICTION OF LONG PIPES.

Gallons per Minute.	DIAMETER OF THE PIPE IN INCHES.										
	1	1½	2	2½	3	3½	4	5	6	7	8	9
	HEAD OF WATER IN FEET.										
320	421·3	55·5	13·16	4·315	1·734	·802	·4115	·13486	·054190	·02507	·012861	·00713
330	448·1	59·0	14·00	4·589	1·844	·853	·4376	·14342	·057630	·02639	·013677	·00759
340	475·6	62·6	14·87	4·871	1·958	·905	·4645	·15224	·061175	·02831	·014519	·00805
350	501·0	66·3	15·75	5·162	2·075	·959	·4923	·16133	·064827	·03000	·015386	·00853
360	533·3	70·2	16·66	5·461	2·196	1·015	·5248	·17068	·068584	·03173	·016277	·00903
370	563·3	74·1	17·60	5·769	2·336	1·072	·5502	·18029	·072447	·03352	·017194	·00954
380	594·2	78·2	18·57	6·085	2·446	1·131	·5803	·19017	·076416	·03536	·018136	·01006
390	625·8	82·4	19·56	6·409	2·576	1·191	·6112	·20031	·080491	·03724	·019103	·01060
400	658·4	86·7	20·57	6·742	2·710	1·253	·6430	·21072	·084672	·03918	·020096	·01115
410	691·7	91·0	21·61	7·083	2·847	1·317	·6755	·22138	·088958	·04116	·021116	·01171
420	725·8	95·5	22·68	7·433	2·988	1·382	·7089	·23231	·093350	·04320	·022155	·01229
430	760·8	100·1	23·8	7·79	3·13	1·448	·743	·2435	·09784	·04528	·023225	·01288
440	796·6	104·9	24·8	8·15	3·27	1·516	·778	·2549	·10245	·04741	·024316	·01349
450	833·2	109·7	26·0	8·53	3·43	1·586	·813	·2666	·10716	·04959	·025434	·01411
460	870·7	114·6	27·2	8·91	3·58	1·657	·850	·2786	·11197	·05182	·026576	·01474
470	909·0	119·7	28·4	9·30	3·74	1·730	·887	·2909	·11690	·05409	·027745	·01539
480	948·0	124·8	29·6	9·70	3·90	1·805	·925	·3034	·12192	·05642	·028938	·01645
490	988·0	130·1	30·8	10·11	4·06	1·881	·964	·3162	·12706	·05880	·030156	·01675
500	1028·7	135·4	32·1	10·53	4·23	1·958	1·004	·3292	·13230	·06122	·031400	·01742
520	1112·7	146·5	34·7	11·39	4·58	2·118	1·086	·3561	·14309	·06622	·033962	·01884
540	1200·0	158·0	37·5	12·28	4·93	2·284	1·171	·3840	·15431	·07141	·036624	·02032
560	1290·4	169·9	40·3	13·21	5·31	2·457	1·260	·4130	·16595	·07680	·039388	·02185

HYDRAULIC TABLE 3—continued.

Gallons per Minute	1	1½	2	2½	3	3½	4	5	6	7	8	9
						Diameter of the Pipe in Inches.						
						Head of Water in Feet.						
580	1384·2	182·2	43·2	14·17	5·69	2·635	1·351	·4430	·17802	·08238	·042251	·02344
600	1481·4	195·0	46·3	15·17	6·09	2·820	1·446	·4741	·19051	·08816	·045216	·02509
620	1581·8	208·3	49·4	16·19	6·51	3·011	1·544	·5062	·20342	·09413	·048280	·02679
640	1685·5	222·0	52·6	17·26	6·93	3·209	1·646	·5394	·21676	·10311	·051445	·02854
660	1792·5	236·0	56·0	18·35	7·37	3·412	1·750	·5736	·23051	·10667	·054711	·03036
680	1902·7	250·5	59·4	19·48	7·83	3·622	1·858	·6089	·24470	·11324	·058077	·03222
700	2016·3	265·5	63·0	20·64	8·30	3·839	1·969	·6453	·25930	·12000	·061544	·03415
720	2133·2	280·9	66·6	21·84	8·78	4·061	2·099	·6827	·27433	·12695	·065111	·03613
740	2253·3	296·7	70·4	23·07	9·44	4·290	2·200	·7211	·28979	·13410	·068778	·03816
760	2376·8	313·0	74·2	24·34	9·78	4·525	2·321	·7606	·30566	·14145	·072546	·04025
780	2503·5	329·6	78·2	25·63	10·30	4·766	2·445	·8012	·32196	·14899	·076415	·04240
800	2633·6	346·8	82·3	26·96	10·84	5·014	2·572	·8428	·33868	·15673	·080384	·04460
820	2766·9	364·3	86·4	28·33	11·39	5·268	2·702	·8855	·35583	·16467	·084464	·04686
840	2903·5	382·3	90·7	29·73	11·95	5·528	2·835	·9292	·37340	·17280	·088623	·04918
860	3043·4	400·7	95·5	31·16	12·52	5·794	2·972	·9740	·39139	·18112	·092893	·05155
880	3186·6	419·6	99·5	32·63	13·11	6·067	3·112	1·0298	·40981	·18965	·097264	·05397
900	3333·1	438·9	104·1	34·13	13·72	6·346	3·255	1·0667	·42865	·19836	·101736	·05645
920	3482·9	458·6	108·8	35·66	14·38	6·631	3·401	1·1147	·44791	·20728	·106307	·05899
940	3636·0	478·8	113·6	37·23	14·96	6·923	3·551	1·1637	·46760	·21639	·110980	·06158
960	3792·4	499·4	118·5	38·83	15·61	7·220	3·703	1·2137	·48771	·22569	·115752	·06423
980	3952·0	520·4	123·5	40·47	16·26	7·524	3·859	1·2648	·50824	·23520	·120626	·06693
1000	4115·0	541·9	128·6	42·14	16·94	7·835	4·019	1·3170	·52920	·24490	·125600	·06970

HYDRAULIC TABLE 3—continued.

HEAD FOR FRICTION OF LONG PIPES.

DIAMETER OF THE PIPE IN INCHES.

HEAD OF WATER IN FEET.

Gallons per Minute	10	12	14	15	16	18	20	21	24
100	·000411	·0001165	·0000765	·0000541	·0000392	·0000217	·0000128	·0000100	·00000516
110	·000497	·0002200	·0000925	·0000655	·0000474	·0000263	·0000155	·0000121	·00000625
120	·000592	·0002238	·0001101	·0000780	·0000565	·0000313	·0000185	·0000145	·00000744
130	·000695	·0002279	·0001293	·0000915	·0000663	·0000368	·0000217	·0000170	·00000873
140	·000806	·0000324	·0001499	·0001062	·0000769	·0000426	·0000252	·0000197	·00001012
150	·000925	·0000372	·0001721	·0001219	·0000883	·0000490	·0000289	·0000226	·00001162
160	·001053	·0000423	·0001958	·0001387	·0001004	·0000557	·0000329	·0000257	·00001323
170	·001189	·0000477	·0002211	·0001566	·0001134	·0000629	·0000371	·0000291	·00001493
180	·001333	·0000535	·0002479	·0001755	·0001270	·0000705	·0000416	·0000326	·00001674
190	·001485	·0000597	·0002762	·0001956	·0001416	·0000786	·0000464	·0000363	·00001865
200	·001646	·0000661	·0003060	·0002167	·0001569	·0000871	·0000514	·0000403	·00002067
210	·001814	·0000729	·0003374	·0002389	·0001730	·0000960	·0000567	·0000444	·00002279
220	·001991	·0000800	·0003703	·0002622	·0001899	·0001054	·0000622	·0000487	·00002501
230	·002176	·0000874	·0004047	·0002866	·0002076	·0001152	·0000680	·0000533	·00002733
240	·002370	·0000952	·0004407	·0003121	·0002260	·0001254	·0000740	·0000580	·00002977
250	·002572	·001033	·0004782	·0003387	·0002452	·0001361	·0000803	·0000629	·00003231
260	·002781	·001117	·0005172	·0003662	·0002653	·0001472	·0000869	·0000681	·00003493
270	·003000	·001205	·0005578	·0003950	·0002861	·0001587	·0000937	·0000734	·00003767
280	·003226	·001295	·0005998	·0004248	·0003076	·0001707	·0001008	·0000789	·00004051
290	·003460	·001390	·0006435	·0004557	·0003300	·0001831	·0001081	·0000847	·00004346
300	·003703	·001488	·0006886	·0004877	·0003532	·0001960	·0001157	·0000906	·00004651
310	·003954	·001589	·0007353	·0005207	·0003771	·0002093	·0001235	·0000968	·00004966

HEAD FOR FRICTION OF LONG PIPES. 13

HYDRAULIC TABLE 3—continued.

DIAMETER OF THE PIPE IN INCHES.

HEAD OF WATER IN FEET.

Gallons per Minute	10	12	14	15	16	18	20	21	24
320	·004213	·001693	·0007832	·0005549	·0004018	·0002230	·0001316	·0001032	·00005292
330	·004481	·001800	·0008332	·0005901	·0004273	·0002371	·0001400	·0001097	·00005628
340	·004757	·001911	·0008845	·0006264	·0004536	·0002517	·0001486	·0001164	·00005974
350	·005041	·002026	·0009373	·0006638	·0004807	·0002678	·0001575	·0001234	·00006331
360	·005333	·002142	·0009916	·0007023	·0005082	·0002822	·0001666	·0001305	·00006697
370	·005633	·002264	·001047	·0007418	·0005372	·0002981	·0001760	·0001379	·00007075
380	·005942	·002388	·0011048	·0007825	·0005667	·0003145	·0001856	·0001454	·00007462
390	·006259	·002515	·001638	·0008242	·0005969	·0003312	·0001956	·0001532	·00007860
400	·006584	·002646	·0012242	·0008670	·0006279	·0003484	·0002057	·0001612	·00008269
410	·006917	·002780	·0012862	·0009109	·0006597	·0003661	·0002170	·0001693	·00008687
420	·007259	·002917	·0013497	·0009559	·0006923	·0003841	·0002268	·0001777	·00009116
430	·00760	·003305	·001414	·001002	·000725	·000402	·000237	·000186	·0000855
440	·00796	·003320	·001481	·001049	·000759	·000421	·000248	·000195	·0001000
450	·00833	·003334	·001549	·001097	·000794	·000441	·000260	·000204	·0001046
460	·00870	·003349	·001619	·001146	·000830	·000460	·000272	·000213	·0001093
470	·00909	·003365	·001690	·001197	·000866	·000481	·000284	·000222	·0001141
480	·00948	·003381	·001762	·001248	·000904	·000501	·000296	·000232	·0001190
490	·00988	·003397	·001837	·001301	·000942	·000522	·000308	·000241	·0001245
500	·01028	·003413	·001912	·001354	·000981	·000544	·000321	·000251	·0001292
520	·01112	·003447	·002069	·001464	·001061	·000588	·000347	·000272	·0001397
540	·01200	·003482	·002231	·001580	·001144	·000635	·000374	·000293	·0001507
560	·01290	·003518	·002399	·001699	·001230	·000683	·000403	·000315	·0001620

HYDRAULIC TABLE 3—continued.

DIAMETER OF THE PIPE IN INCHES.

HEAD OF WATER IN FEET.

Gallons per Minute.	10	12	14	15	16	18	20	21	24
580	·01384	·00536	·002574	·001823	·001320	·000732	·000432	·000338	·0001738
600	·01481	·00595	·002754	·001950	·001412	·000784	·000462	·000362	·0001860
620	·01581	·00635	·002941	·002083	·001508	·000837	·000494	·000387	·0001986
640	·01685	·00677	·003134	·002219	·001607	·000892	·000526	·000412	·0002116
660	·01792	·00720	·003333	·002360	·001709	·000948	·000560	·000438	·0002251
680	**·01902**	·00764	·003538	·002505	·001814	·001007	·000594	·000465	·0002389
700	**·02016**	·00810	·003749	·002655	·001923	·001071	·000630	·000493	·0002532
720	**·02133**	·00856	·003966	·002809	·002032	·001129	·000666	·000523	·0002679
740	**·02253**	·00905	·004190	·002967	·002151	·001192	·000704	·000551	·0002830
760	**·02376**	·00955	·004419	·003130	·002266	·001258	·000742	·000581	**·0002985**
780	·02503	·01006	·004635	·003297	·002387	·001325	·000782	·000613	·0003144
800	·02633	·01058	·004897	·003468	·002511	·001393	·000823	·000644	·0003307
820	·02767	·01112	·005144	·003643	·002638	·001464	·000868	·000677	·0003475
840	·02903	·01166	·005398	·003823	·002769	·001536	·000907	·000710	·0003646
860	·03043	·01223	·005659	·004008	·002902	·001610	·000951	·000745	·0003822
880	·03186	·01280	·005925	·004196	·003038	·001686	·000995	·000780	·0004002
900	·03333	·01339	·006197	·004389	·003178	·001764	·001041	·000816	·0004186
920	·03483	·01399	·006476	·004586	·003321	·001843	·001088	·000852	·0004374
940	**·03636**	·01461	·006760	·004788	·003467	·001924	·001136	·000890	·0004566
950	**·03792**	·01524	·007051	·004994	·003616	·002007	·001184	·000928	·0004763
980	**·03952**	·01588	·007348	·005204	·003769	·002091	·001235	·000967	·0004982
1000	**·04115**	·01653	·007651	·005419	·003924	·002178	·001286	·001007	·0005168

HEAD FOR FRICTION OF LONG PIPES.   15

DIAMETER OF THE PIPE IN INCHES.

HEAD OF WATER IN FEET.

Gallons per Minute.	5	6	7	8	9	10	12
2,000	5·2	2·11	·97	·50	·27	·164	·066
3,000	11·8	4·76	2·20	1·13	·62	·370	·148
4,000	21·0	8·46	3·91	2·00	1·11	·658	·264
5,000	32·9	13·23	6·12	3·14	1·74	1·02	·413
6,000	47·4	19·05	8·81	4·52	2·50	1·48	·595
7,000	64·5	25·93	12·00	6·15	3·41	2·01	·810
8,000	84·2	33·86	15·67	8·03	4·46	2·63	1·05
9,000	106·6	42·86	19·83	10·17	5·64	3·33	1·33
10,000	131·7	52·92	24·49	12·56	6·97	4·11	1·65
20,000	526·8	211·68	97·96	50·24	27·88	16·46	6·61

DIAMETER OF THE PIPE IN INCHES.

HEAD OF WATER IN FEET.

Gallons per Minute.	14	15	16	18	20	21	24
2,000	·0306	·0216	·0156	·0087	·0051	·0040	·0020
3,000	·0688	·0487	·0353	·0196	·0115	·0090	·0046
4,000	·122	·0867	·0627	·0348	·0205	·0161	·0082
5,000	·191	·135	·0981	·0544	·0321	·0251	·0129
6,000	·275	·195	·141	·0784	·0462	·0362	·0186
7,000	·374	·265	·192	·107	·0630	·0493	·0253
8,000	·480	·346	·251	·139	·0823	·0644	·0330
9,000	·619	·438	·317	·176	·104	·0816	·0418
10,000	·765	·541	·392	·217	·128	·100	·0516
20,000	3·06	2·16	1·56	·871	·514	·403	·206
30,000	6·88	4·87	3·53	1·96	1·15	·906	·465
40,000	12·24	8·67	6·27	3·48	2·05	1·61	·826
50,000	19·12	13·54	9·81	5·44	3·21	2·51	1·29
60,000	27·54	19·50	14·12	7·84	4·62	3·62	1·86
70,000	37·49	26·55	19·23	10·71	6·30	4·93	2·53
80,000	48·97	34·68	25·11	13·93	8·23	6·44	3·30

desired, which will only show that the exact diameter is an odd size between the standard ones in the Table. But by the former rule in (6), this can be easily checked; thus, in our case, the true head for an 8-inch pipe would be $\cdot 01286 \times 1600 = 20 \cdot 57$ feet instead of 20 feet; but, of course, in most cases 8 inches is near enough for practice.

(8.) 3rd. To find G, having H, L, and d given. Divide the given head of water in feet by the given length in yards, and the nearest number thereto in the Table, under the given diameter, will be found opposite the required number of gallons. Thus, to find the discharge of a 7-inch pipe 3797 yards long with 45 feet head, see (5), we have $\frac{45}{3797} = \cdot 01185$; and looking for this under 7 inches diameter, we find it opposite 220 gallons, the discharge sought. Again, for the discharge of a 10-inch pipe 3000 yards long with 40 feet head, we have $\frac{40}{3000} = \cdot 01333$; and the nearest number to that we find to be $\cdot 01384$ opposite 580 gallons, the discharge sought.

(9.) 4th. To find L, having H, G, and d given. Divide the given head by the head for one yard found in the Table under the given diameter, and opposite the given number of gallons, and the result is the required length. Thus, to determine the length of 4-inch pipe to consume 12 feet head with 130 gallons per minute, we find under 4 inches and opposite 130 gallons $\cdot 0679$ the head for one yard, and hence $\frac{12}{\cdot 0679} = 176$ yards, the length sought.

(10.) To avoid a needless extension of the Table, we have given only the principal numbers from 1 to 90, and from 1000 to 100,000 gallons, leaving the intervening numbers to be supplied from the body of the general Table. In order to do this, it should be observed that the head varies as the square of the discharge, so that, for instance, ten times any given discharge will require 100 times the head, &c., &c. Thus, with 100 gallons, the Table shows that a 5-inch pipe requires $\cdot 01317$ foot head per yard, then with 1000 gallons the head would be $\cdot 01317 \times 100 = 1\cdot 317$ foot; and with 10 gallons $\frac{\cdot 01317}{100} =$ ·0001317 foot. The application of this principle to any case in practice is very simple: say we require the head for 33 gallons with a 2½-inch pipe 600 yards long. Not finding 33 gallons in the Table, we take 330, the head for which is 4·589, therefore for 33 gallons it will be $\frac{4\cdot 589}{100} = \cdot 04589$. This may be checked by the skeleton Table, which shows that 30 gallons require ·03792, and 40 gallons ·06742 foot; so that ·04589 looks about right for 33 gallons. Then the head required in our case is ·04589 × 600 = 27·534 feet.

Again, say we required the head for 2800 gallons with a 15-inch pipe 500 yards long. Here we must take the head for 280 gallons from the Table, which is ·0004248: for 2800 gallons, therefore, or 10 times the quantity, we should have ·0004248 × 100 = ·04248 foot. Checking this by the skeleton Table we find ·0487 foot for 3000 gallons, showing that ·04248 foot for 2800 gallons is about right. Hence the head sought is, in our case, ·04248 × 500 = 21·24 feet.

The same principle may be applied when the discharge is the unknown quantity; thus, to find the discharge of a 2½-inch pipe, 700 yards long with 17 feet head, we have $\frac{17}{700} = \cdot 02428$, which, by the skeleton Table, is somewhere between 20 and 30 gallons: now, looking in the body of the Table between 200 and 300 gallons for the same figures (neglecting altogether for the moment the position of the decimal place) we find that the nearest to 2428 is 2427, which is opposite 240 gallons; 24 gallons is therefore the true discharge. Again, to find the discharge of a pipe 1½-inch diameter, 200 yards long, with 4·5 feet head, we have $\frac{4\cdot 5}{200} = \cdot 0225$, which, by Table, is between 6 and 7 gallons; now, looking between 600 and 700 gallons, we find the nearest to be 222 opposite 640 gallons, and as we know that the true discharge is between 6 and 7 gallons, we infer that the exact quantity is 6·4 gallons, &c., &c.

(11.) The 3rd illustration in (8) for finding G may be extended so as to give a useful general view of the discharge of different sized pipes with the same length and head. Thus, we found the tabular number for 3000 yards long and 40 feet head to be $\frac{40}{3000}$ = ·01333, and looking for this successively under different diameters we find that

A 6-inch pipe discharges 160 gallons per minute
,, 7 ,, ,, 235 ,, ,,
,, 8 ,, ,, 330 ,, ,,
,, 9 ,, ,, 440 ,, ,,
,, 10 ,, ,, 580 ,, ,,
,, 12 ,, ,, 900 ,, ,, &c.

(12.) "*Head for Velocity of Entry.*"—To the head thus found by the preceding rules and Table, that due to velocity of entry has in all cases to be added, as explained in (4). When the pipe is of the common form, with square edges, as in Figs. 3 and 4, Table 2 gives the head for velocity direct. For very long pipes this is so small in proportion to the head due to friction, that it may in such cases be neglected, and we have omitted it for that reason in the preceding illustrations; thus, we found in (5) and in (6) that with 320 gallons, by an 8-inch pipe 3457 yards long, the head due to friction alone was 44·46 feet. By Table 2 it will be seen that the head for velocity at entry is rather less than 2 inches, so that in such a case it may be neglected. But when a pipe is very short, the head due to velocity may be much greater than that due to friction, and the most serious errors may be made by neglecting it. Say we had an 18-inch pipe, 20 yards long, discharging 3000 gallons. By Table 3 the friction is ·0196 × 20 = ·392 foot; and the head due to velocity by Table 2 is 6 inches, or ·5 foot, being more than that due to friction; so that the total head is ·392 + ·5 = ·892 foot.

(13.) When, with a very short pipe, the head is given and the discharge has to be calculated, the case does not admit of a simple direct solution, because we cannot tell beforehand in what proportions the total head at disposal has to be divided between overcoming friction and generating velocity. We must for such cases, apply a useful general law (27), which may be stated as follows :—" *The discharge by any pipe, or series of pipes, is proportional to the square root of the head;*" and conversely, " *The head is proportional to the square of the discharge ;*" and these laws are **true** in pipes with bends, jets, contractions, &c. Thus, say we require the discharge of a 12-inch **pipe** 5 yards long with 10 feet head. Assume a discharge, it is unimportant whether the assumed discharge **is near** the true quantity or not, **or whether it is too much** or **too little**. Say, in our case, we take it at 1000 gallons per minute, then by Table 3 the head for friction is ·01653 × 5 = ·08265 foot, and the head for velocity is, by Table 2, about 4 inches, or ·333 foot, making a total of ·08265 + ·333 = ·41565 foot, instead of 10 feet, the head at disposal. Then applying the law just given, we have $\frac{1000 \times \sqrt{10}}{\sqrt{\cdot 41565}} = \frac{1000 \times 3 \cdot 162}{\cdot 6447} = 4905$ gallons. Now, if in this case the head due to velocity had been neglected, the discharge by Table 3 would be $\frac{10}{5} = 2 \cdot 0 = 11{,}000$ gallons, which is more **than double the true discharge.** The **Table 2** gives the **greatest possible facility for making** the calculations of head due to velocity, which should never be overlooked **in cases** where the **pipe is short.**

(14.) "*Loss of Head by Bends.*"—There is another source of loss of head in pipes—namely, **change of direction, or** bends. The best formula for calculating **this loss** is that of Weisbach, which may be modified into the following :—

$$H = \left\{ \cdot 131 + (1 \cdot 847 \times \left(\frac{r}{R}\right)^{\frac{7}{2}} \right\} \times \frac{V^2 \times \phi}{960},$$

$$\text{and } V^2 = \frac{960 \times H}{\phi \times \left\{ \cdot 131 + (1 \cdot 847 \times \left(\frac{r}{R}\right)^{\frac{7}{2}} \right\}};$$

LOSS OF HEAD BY BENDS.

In which H = the head due to change of direction, in inches.
    r = radius of the bore of the pipe, in inches.
    R = radius of the centre line of the bend, in inches.
    φ = angle of bend, in degrees.
    V = velocity of discharge, in feet per second.

Thus, say we require the loss of head by a bend of 9 inches radius in a 6-inch pipe, discharging 800 gallons per minute, with an angle of 55°. A 6-inch pipe containing roughly $\frac{6^2}{30} = 1\cdot 2$ gallon per foot run, the velocity of discharge will be $\frac{800}{1\cdot 2 \times 60}$ = 11·1 feet per second. To find $\left(\frac{r}{R}\right)^{\frac{7}{2}}$, or in our case $\left(\frac{3}{9}\right)^{\frac{7}{2}}$, we have $\frac{3}{9} = \cdot 3333$.

Then the log. of ·3333 = $\overline{1}\cdot 522835$
$$\frac{7}{2)\overline{4}\cdot 659845}$$
$$\overline{2}\cdot 329922 = \cdot 02137 = \left(\frac{3}{9}\right)^{\frac{7}{2}}$$

Then $\{\cdot 131 + (1\cdot 847 \times \cdot 02137\} \times \frac{11\cdot 1^2 \times 55}{960} = 1\cdot 2$ inch, the head required.

Table 4 has been calculated by the second formula. The first part is adapted to bends of the radius usually met with in practice; this may vary slightly with different makers, but not so much as to affect the result seriously. Fig. 6 gives the proportions of the 8-inch bend as an illustration. The second part of the Table gives the loss by *quick* bends of the proportions given by Fig 7, which are sometimes necessary in special cases; they are commonly named "elbows."

Table 4 requires but little explanation; it shows, for instance, that an ordinary 8-inch bend, with 18 inches radius, consumes 3 inches head when passing 1970 gallons per minute; but a quick 8-inch bend with 6 inches radius consumes 12 inches

LOSS OF HEAD BY BENDS. 21

TABLE 4.—TABLE for BENDS in WATER PIPES, showing the LOSS of HEAD DUE to CHANGE of DIRECTION by ONE BEND of 90°.

Diameter of the Pipe in Inches.	Radius of Centre line of Bend in Inches.	HEAD OF WATER IN INCHES LOST BY ONE BEND OF 90°.														
		¼	½	¾	1	1½	2	3	4	5	6	9	12	18	24
		GALLONS DISCHARGED PER MINUTE.														
2	12	25	36	51	63	73	81	103	126	146	163	179	219	252	309	358
3	12	58	83	117	144	166	203	235	288	332	371	407	498	576	705	814
4	12	102	145	205	252	291	356	411	504	582	650	713	873	1008	1233	1426
5	18	162	229	324	396	458	561	648	793	916	1024	1122	1374	1586	1944	2244
6	18	232	328	464	568	656	803	928	1136	1312	1467	1607	1968	2272	2784	3124
7	18	309	437	618	757	874	1070	1236	1514	1748	1954	2141	2622	3028	3708	4282
8	18	402	568	804	985	1137	1393	1608	1970	2274	2542	2786	3411	3940	4824	5572
9	18	501	709	1002	1228	1418	1737	2005	2456	2836	3170	3474	4254	4912	6015	6948
10	18	606	857	1212	1484	1714	2100	2424	2968	3428	3832	4199	5142	5936	7272	8398
12	21	866	1225	1733	2122	2451	3003	3466	4245	4902	5480	6005	7353	8490	10398	12010
15	24	1317	1864	2635	3228	3728	4567	5271	6457	7456	8336	9131	11184	12914	15813	18268
18	27	1857	2626	3714	4549	5253	6435	7428	9098	10505	11745	12870	15759	18196	22284	25740
21	30	2467	3490	4935	6044	6980	8550	9870	12089	13960	15607	17100	20940	24178	29610	34200
24	33	3165	4477	6330	7754	8954	10968	12661	15508	17908	20021	21937	26862	31016	37983	43870

TABLE FOR QUICK BENDS.

2	3	23	32	46	56	65	79	92	112	130	145	159	195	214	276	318
3	3½	44	63	89	109	126	154	178	218	252	282	309	378	436	534	618
4	4	69	98	139	170	197	241	278	341	394	440	480	591	682	834	966
5	4½	96	136	172	236	272	333	385	472	544	608	666	816	944	1155	1332
6	5	128	181	256	314	362	443	512	629	724	809	886	1086	1258	1536	1774
7	5½	161	229	322	396	458	561	645	793	917	1024	1122	1374	1586	1935	2244
8	6	199	281	398	487	563	689	796	975	1126	1259	1379	1689	1950	2388	2758

head when passing nearly the same **quantity, or 1950** gallons, and these, it should be observed, **are the** heads due simply to *change of direction*, **and do not include the** head due to velocity or to friction. Thus, for instance, if the quick **8-inch bend had a length of one yard, the head for** friction by Table 3 (say for **2000 gallons) would be** ·5 foot, and the head for velocity at entry by the rule in (3), namely $\left(\dfrac{G}{d^2 \times 13}\right)^2 = H$ is $\left(\dfrac{1950}{8^2 \times 13}\right)^2 = 5\cdot 48$ feet. Thus we have a total for such a bend of

    1·0 feet for change of direction,
    0·5 ,, for friction,
    5·48 ,, for velocity at entry,
    ―――
    6·98 ,, total.

Again, in a 6-inch pipe carrying 800 gallons, the Table shows that each **common bend causes a loss of** 1½ **inches** head, and each quick **bend a loss of 5 inches, &c.** The Table is arranged for bends **of 90°, or quarter bends, as they are technically named,** but **it is applicable to any other angle, for the loss of head** is simply proportional **to the angle, the radius being the same; thus, a** half-quarter bend **of 45°, or one-eighth part of a** circle, **consumes** half the head **of a bend of 90°, and a bend of** 180°, **or half a** circle, takes **double, &c., &c.**

(15.) "*Discharge of Compound Water-mains.*"—When a long main is composed of pipes of different **sizes, as** is very frequently the case, the head for each must be **separately** calculated, and the sum total taken. Thus, if we required 300 gallons **per minute** through a main 1200 **yards long, composed of 800 yards of 7-inch,** 300 yards of 6-inch, **and 100 yards of** 5-inch pipe, the head would be—

        By Table 3.
 300 gallons 7-inch = ·022 × 800 = 17·6 feet head
  ,,   ,,   6 ,, = ·0476 × 300 = 14·28   ,,
  ,,   ,,   5 ,, = ·1185 × 100 = 11·85   ,,
             ―――
             43·73 total.

If there were bends in the pipes we must add the head for

them from Table 4, but it will be found, as in the case of head for velocity, see (12), that with long mains the effect of bends is very small. Say we had

4	common bends	in the	7-inch,	each	$\frac{1}{8}$-inch head	=	$\frac{1}{2}$	inch
3	quick	,,	7	,,	$\frac{1}{2}$,,	=	$1\frac{1}{2}$	,,
2	common	,,	6	,,	$\frac{1}{4}$,,	=	$\frac{1}{2}$	,,
2	quick	,,	6	,,	$\frac{3}{4}$,,	=	$1\frac{1}{2}$	,,
4	common	,,	5	,,	$\frac{1}{2}$,,	=	2	,,
3	quick	,,	5	,,	$1\frac{1}{2}$,,	=	$4\frac{1}{2}$	,,

Total $10\frac{1}{2}$ inches.

Thus, even for such a large number of bends, the loss of head is only $10\frac{1}{2}$ inches, or ·875 of a foot; so that the total loss is $43 \cdot 73 + \cdot 875 = 44 \cdot 605$ feet.

(16.) When, with such a series of pipes the head is given, and the discharge has to be determined, the case does not admit of a direct solution, because we cannot tell beforehand in what proportions the given head must be divided among the different pipes. We must in that case follow the course explained in (13): thus, say we required the discharge with 30 feet head by a main 2000 yards long, composed of 1200 yards of 8-inch pipe with four common bends in it; 700 yards of 6-inch pipe and three bends; and 100 yards of 5-inch pipe, with two common and two quick bends. The first thing to be done is to assume a discharge, and calculate the head for that, as was done in the last example; it is unimportant whether the assumed discharge is near the true quantity or not. Say in our case we take it at 400 gallons. Then

By Table 3.   Length.   Feet.
400 gallons 8-inch pipe = ·02 × 1200 = 24·0 head
  ,,   6   ,,   = ·085 × 700 = 59·5 ,,
  ,,   5   ,,   = ·21 × 100 = 21·0 ,,

Carried forward .. 104·5

24   EFFECT OF CONTOUR OF SECTION ON WATER-MAINS.

$$\begin{array}{llllll}
 & & & & \text{Brought forward} & .. \ 104 \cdot 5 \text{ feet} \\
 & & & \text{Inch.} & \text{Inch.} & \text{Inch.} \\
4 \text{ common bends in} & 8 \text{ each} & \tfrac{1}{8} \times 4 = & \tfrac{1}{2} \text{ head} \\
3 \quad ,, & ,, & 6 \ ,, & \tfrac{1}{2} \times 3 = & 1\tfrac{1}{2} \ ,, \\
2 \quad ,, & ,, & 5 \ ,, & \tfrac{3}{4} \times 2 = & 1\tfrac{1}{2} \ ,, \\
2 \text{ quick} & ,, & 5 \ ,, & 3 \times 2 = & 6 \ ,, \\
 & & & & \overline{9\tfrac{1}{2}} = & \cdot 8 \text{ foot} \\
 & & & & \text{Total} & \overline{105 \cdot 3 \text{ feet.}}
\end{array}$$

Thus we find that for 400 gallons we require 105·3 feet head instead of 30 feet, the head given; then by the rule in (13) we have $\dfrac{\sqrt{30} \times 400}{\sqrt{105 \cdot 3}}$ or $\dfrac{5 \cdot 447 \times 400}{10 \cdot 26} = 213$ gallons, the real discharge sought. Further illustrations will be found in Chapter II.

(17.) "*Effect of Contour of Section.*"—The contour of the section of the line of pipes is a matter of some importance. The best condition, when the pipe is of uniform diameter from end to end, is, of course, a uniform slope throughout. This, however, can rarely be obtained, the pipe having to follow the contour of the ground, as in Fig 9. If a number of open-topped pipes were inserted anywhere along the main, as at A, B, C, D, &c., the water would rise in them to the level of the oblique line J K, which in the case of a pipe of the same bore from end to end, would be a straight line as shown; this line is termed the *hydraulic mean gradient.* Now, the vertical distance from any point in that line (say the top of E) to the level line K M, will give the head for friction between E and K, and the vertical distance from the same point to the level line J L will give the friction between E and J: we have here supposed, of course, that the figure is correctly drawn to scale.

(18.) When, as in Fig. 11, the pipes are of different diameters, then each would have its own gradient, showing at every point the loss of head due to that particular pipe as in the figure. No loss of effect will arise from the pipe following the section of the ground, *so long as the contour of the pipe does not anywhere along the line rise above the hydraulic mean gradient.* Thus, in

Fig. 9, where the ground is much broken, but does not anywhere rise above the gradient, the discharge will be the same as by a pipe with a uniform slope.

(19.) But if, as in Fig. 10, a hill, as at B, rises higher than the gradient, then the pipe from C to D will be in a state of partial vacuum, air will be given out by the water, and will accumulate at the summit, and being driven forward by the water from C to B, will remain permanently in the pipe from B to G, occupying the upper part of the pipe while the water trickles down the lower part as in a trough or open channel, and the vertical head from B to G is lost, the hydraulic gradient being now from A to B, from B to G, and from G to F, this last being parallel to that from A to B, or at the same angle with the horizon. The discharge at F will therefore be, not the amount due to the head E, F on the length A, F, but that due to the head E, B on the length A, B.

(20.) In this case the size of the pipe should not be uniform from end to end: from A to B it should be of large diameter, so as to deliver at B the required quantity with the head E, B; and the pipe from B to F may be of smaller diameter, so as to deliver the same quantity at F with the head H, F. Say we take a case with the length A, F = 5000 yards, and head E, F = 90 feet, and that the length A, B = 2400 yards, and the head E, B = 10 feet, and that 500 gallons were required at F. With uniform slope we should have $\frac{90}{5000}$ = ·018, which, by Table 3, is a 9-inch pipe, or rather less, for a 9-inch pipe would deliver 500 gallons with ·01742 × 5000 = 87·1 feet. But for the delivery at B with 10 feet head, and a 9-inch pipe, we have $\frac{10}{2400}$ = ·004167, which by Table = 245 gallons only, instead of 500; and, of course, this is all we should get at F with such an arrangement, for whatever the size of the rest of the pipe from B to F might be, it could not deliver more than it received by the pipe A, B.

The pipe from A to B should be $\frac{10}{2400}$ = ·004167, by Table 3

= a 12-inch pipe; and the pipe from B to F may be $\frac{80}{2600}$ = ·03077 = an 8-inch pipe by Table. We may check these results thus:—

	By Table 3.	Length.	Head.
12-inch pipe, 500 gallons =	·00413 ×	2400 =	9·912 feet
8 ,, 500 ,, =	·0314 ×	2600 =	81·64 ,,
		Total	91·552

Thus we find the exact head to be a little more than the head at disposal, but in most cases the agreement is near enough for practice.

(21.) When a long main is composed of different sizes of pipes and passes over uneven ground, the best course is to draw the gradients on the section of the pipes so as to see at a glance that none of the hill-tops rise above them. Fig. 11 is a case in which, with a fall of 232 feet, we have a 10-inch main 4000 yards long, an 8-inch main 3000 yards long, and a 6-inch main 2000 yards long. To divide the given fall in the proper proportion between the different pipes and so find the gradients, let us assume that 100 gallons are delivered; then

	By Table 3.	Length.	A.
100 gallons 10-inch =	·000411 ×	4000 =	1·644 feet head
,, 8 ,, =	·001256 ×	3000 =	3·768 ,,
,, 6 ,, =	·005292 ×	2000 =	10·584 ,,
			15·996 total head.

Now, whatever the real head may be, it would have to be divided among the several pipes in the same proportions as for 100 gallons in Col. A, and as the head in our case is $\frac{232}{15·996}$ = 14·504 times the total head for 100 gallons, it follows that the real head for each pipe will be 14·504 times the head for the same pipe in Col. A; thus the true head

E, B for the 10-inch pipe will be	1·644 × 14·504 =	23·84 feet
F, C ,, 8 ,, ,,	3·768 × 14·504 =	54·65 ,,
G, D ,, 6 ,, ,,	10·584 × 14·504 =	153·51 ,,
		232·00

We can now draw the gradients on the section as in Fig. 11, and then if the contour of the ground is below them throughout, all is well.* The discharge at D may be calculated from any one of the pipes; say we take the 8-inch; then $\frac{54\cdot 65}{3000} = \cdot 01822 =$ about 380 gallons by Table 3.

(22.) "*Special Cases.*"—There are many cases for the solution of which no general rules can be given—they require reasoning, with the assistance of rules. The following cases may be useful :—Say that with pipes, arranged as in Fig. 12, we require 50 gallons at B, and 100 gallons at A, and have to determine the sizes of the mains. If we assume 3 inches for E, the head for that size would be $\cdot 0423 \times 160 = 6\cdot 77$ feet above the level at B, and as that point is 8 feet (or $18 - 10$) above the level at C, we have at this last point the head of $6\cdot 77 + 8 = 14\cdot 77$ feet to deliver 50 gallons at B. Now, as A is $25 - 18 = 7$ feet below C, the head on A will be $14\cdot 77 + 7 = 21\cdot 77$ feet, and to find the size of pipe with that head for 100 gallons, we have $\frac{21\cdot 77}{250} = \cdot 0871 =$ a $3\frac{1}{2}$-inch pipe by Table 3. We have now only to fix the size of the pipe D to carry $50 + 100 = 150$ gallons: we found the head at C necessary for the pipes E and F to be $14\cdot 77$ feet, leaving therefore only $18 - 14\cdot 77 = 3\cdot 23$ feet for the friction of D, and from this we find $\frac{3\cdot 23}{300} = \cdot 01077 =$ a 6-inch pipe by Table 3.

(23.) Take another case shown by Fig. 13, and say that we require the head at D to deliver 600 gallons at E by the single and double line of pipes; also to find what proportion of the 600 gallons passes by the two branches A, C, B and A, B. Let us assume that the pipe A, C, B carries 1000 gallons; then the head at A for that quantity would be—

1000 gallons 12-inch pipe = $\cdot 01653 \times 1100 = 18\cdot 18$ feet head
,,   9   ,,   = $\cdot 0697 \times 800 = \underline{55\cdot 76}$,,
                                            $73\cdot 94$,,

---

* The principle of this method of calculating a series of gradients is due to C. E. Amos, Esq., of The Grove, Southwark.

And with that head at A, the pipe A, B would at the same time deliver $\frac{73 \cdot 94}{950}$ = ·0778 = 790 gallons by Table 3; so that the two sets of pipes deliver at B 1790 gallons with a head of 73·94 feet at A, and therefore (13) to deliver the 600 gallons required, would take $\frac{73 \cdot 94 \times 600^2}{1790^2}$ = 8·3 feet. Then, the 12-inch pipe from D to A would require for 600 gallons ·00595 × 1100 = 6·545 feet head, and the 9-inch pipe from B to E, ·02509 × 400 = 10·036 feet; thus the total head at D will be 6·545 + 8·3 + 10·036 = 24·881 feet. The pipe A, C, B will carry $\frac{600 \times 1000}{1790}$ = 336 gallons, therefore the pipe A, B must take the rest, or 264 gallons.

(24.) If the head had been given, and the discharge due thereto had to be determined, we must have calculated the head for an assumed discharge, and then applied the rule in (13) to find the real discharge with the true head. Thus, say that with the same arrangement of pipes, we require the discharge at E with 45 feet head at D. If we assume 600 gallons, we should find 24·881 feet head as in (23); then $\frac{600 \times \sqrt{45}}{\sqrt{24 \cdot 881}}$ or $\frac{600 \times 6 \cdot 708}{4 \cdot 988}$ = 807 gallons, the discharge at E with 45 feet head at D, &c.

(25.) "*Delivery and Suction-pipes to Pumps.*"—In calculating the sizes of pipes to pumps, it should be remembered that the **action of a** pump is intermittent, especially where there is **no** air-vessel to equalize the velocity of supply and discharge. Say we have a single-acting pump 2 feet diameter **and 2 feet stroke**, worked by a crank, &c., making 16 revolutions per minute. The area of the pump **being** 3·1416 feet, we should have 3·1416 × 2 × 16 = 100 gallons discharged per **minute**; but while the bucket is **descending the delivery** is *nothing*, **and it rises** to a **maximum when the bucket is at the** centre of its up-stroke, where it has the velocity of the crank-pin; thus in our case the crank-path being 2 feet diameter, or 6·28 feet circumference, the maximum discharge at that moment is $6·28 \times 16 \times 3·1416 =$ 314 gallons, and the pipes must be calculated for that quantity instead of 100 gallons, the mean discharge. In most cases, an air-vessel is used, which more or less effectively regulates and equalizes the velocity of discharge: where the suction-pipe is a long one, an air-vessel should be provided for that also. Table 5 gives the variation in velocity in different kinds of pumps without air-vessels.

TABLE 5.—Of the VELOCITY of DISCHARGE by PUMPS WITHOUT AIR-VESSELS.

	Velocity of Discharge.			Variation per cent.
	Max.	Mean.	Min.	
One single-acting pump, worked by a crank	314·16	100	000	314·16
Two ditto, worked by cranks at right angles	222·00	100	000	222·00
One double-acting pump	157·08	100	000	157·08
Three-throw single-acting	104·76	100	90·69	14·07
Four single-acting, or two double-acting	111·00	100	78·79	32·21

This Table shows that the common 3-throw pump has a more uniform discharge than any other, the maximum velocity being under 5 per cent. in excess of the mean; an air-vessel is hardly necessary for such a case, in fact large pumps throwing 600 gallons per minute have been worked for many years successfully without any air-vessel.

(26.) "*Service-pipes in Towns.*"—The sizes of street service-pipes for town supplies cannot be calculated by the ordinary rules: we may pursue another method. Certain sizes of lead services varying with the sizes of the houses supplied have been found necessary by experience. For ordinary cases with intermittent supply we may admit that ½-inch pipe will suffice for a house with 6 or 7 rooms, ⅝-inch for 10 rooms, ¾-inch for 16 rooms, and 1-inch for say 30 rooms. The discharging power of long pipes varies, as the 2·5 power of the diameter (28), thus $4^{2\cdot 5}$ = 32, and we shall therefore require 32 1-inch pipes to deliver with the same head and length the same quantity of water as a 4-inch pipe, and we may admit that a 4-inch main would supply 32 1-inch lead services, &c. Table 6 is calculated on these principles.

TABLE 6.—SERVICE MAINS for WATER-SUPPLY in TOWNS.

Diameter of Branch Mains.	Diameter of Lead Services.			
	$\frac{1}{2}$	$\frac{5}{8}$	$\frac{3}{4}$	1
	Number of Houses supplied.			
$1\frac{1}{2}$	15	9	6	3
2	32	18	12	6
$2\frac{1}{2}$	56	32	20	10
3	88	50	32	15
$3\frac{1}{2}$	..	74	47	23
4	..	104	66	32

"*General Laws for Pipes.*"—The following general statement of the laws governing pipe questions may be useful : some of these laws apply strictly only to long mains in which the head due to velocity may be neglected.

(27.) When d and L are constant, the discharge, or G, varies directly as the square root of the head, so that for heads in the ratio 1, 2, 3, the discharge would be in the ratio $\sqrt{1}$, $\sqrt{2}$, and $\sqrt{3}$, or 1, 1·414, and 1·732.

Conversely,—the head is directly as the square of the discharge, so that for discharges in the ratio 1, 2, 3, we require heads in the ratio 1^2, 2^2, 3^2, or 1, 4, 9, &c.

(28.) When H and L are constant, the discharge is directly as the 2·5 power of the diameter; thus with diameters in the ratio 1, 2, 3, the discharge will be in the ratio $1^{2\cdot 5}$, $2^{2\cdot 5}$, and $3^{2\cdot 5}$, or 1, 5·6, and 15·6.

Conversely,—the diameter will vary directly as the 2·5 root of the discharge ; thus for discharges in the ratio 1, 2, 3, the diameter will vary in the ratio $\sqrt[5]{1}$, $\sqrt[5]{2}$, and $\sqrt[5]{3}$, or 1, 1·32, and 1·55, &c.

(29.) When G and L are constant, the head will be *inversely* as the 5th power of the diameter; so that for diameters in the ratio 1, 2, 4, the heads will be in the ratio 4^5, 2^5, and 1^5, or 1024, 32, and 1.

Conversely,—the diameter will be inversely as the 5th root of the head; thus for heads in the ratio 1, 2, 4, the diameters would be in the ratio $\sqrt[5]{4}$, $\sqrt[5]{2}$, and $\sqrt[5]{1}$, or 1·32, 1·15, and 1·0, &c.

(30.) When H and d are constant, the discharge will be inversely as the square root of the length; thus for lengths in the ratio 1, 2, 4, the discharge would be in the ratio $\sqrt{4}$, $\sqrt{2}$, and $\sqrt{1}$, or 2·0, 1·414, and 1·0, &c.

Conversely,—the length varies inversely as the square of the discharge; thus for discharges in the ratio 1, 2, 4, the lengths would be in the ratio 4^2, 2^2, and 1^2, or 16, 4, and 1, &c.

(31.) When G and d are constant, the head is directly and simply as the length; thus for lengths in the ratio 1, 2, 3, the heads would also be in the ratio 1, 2, 3, &c.

(32.) "*Head for very Low Velocities.*"—Table 3 gives the greatest possible facility for the calculation of pipe questions, as may be seen by the examples we have given, and for all ordinary cases the results are correct; but for very small velocities with low heads, say under one foot, &c., experiment has shown that the discharges are less than that Table would give, and for such cases Prony's more difficult and laborious rule seems to give the most correct results. The following rule is based on that of Prony:—

Let d = diameter of the pipe in inches.
H = head of water in inches.
L = length of pipe in feet.
G = gallons per minute.

Then
$$\left(16\cdot353 \times \frac{H \times d}{L} + \cdot00665\right)^{\frac{1}{2}} - \cdot0816) \times d^2 \times 2\cdot04 = G.$$

PIPES—SPECIAL RULES FOR LOW VELOCITIES.

Thus, say we required the discharge by a 12-inch pipe 3000 feet long with 36 inches head: then

$$\left(16{\cdot}353 \times \frac{36 \times 12}{3000} + {\cdot}00665\right)^{\frac{1}{2}} - {\cdot}0816\right) \times 144 \times 2{\cdot}04 = 427{\cdot}4 \text{ gallons.}$$

We may compare this result with that by Table 3, or rather by the rule $\left(\frac{(3d)^5 \times H}{L}\right)^{\frac{1}{2}} = G$, given in (5), by which the discharge comes out 426 gallons, or practically the same as by Prony's rule. With a very small head, however, the two rules do not agree; thus, with **only one** inch head, this same pipe gives 54·87 gallons by Prony's rule, whereas the other rule gives 70·98 gallons, or 29 per cent. more. With a large head, on the contrary, Prony's rule gives a rather larger discharge than the other. The general comparison of the two rules may be shown by the case of a 10-inch pipe, 1000 yards long, the calculated discharge of which, with different heads, is given by the following Table:—

	Head of Water.					
	in. 1	ins. 4	ft. ins. 1 4	ft. ins. 5 4	ft. ins. 21 4	ft. ins. 85 4
	Discharge in Gallons per Minute.					
By the Rule in (5) ..	45	90	180	360	720	1440
By Prony's Rule ..	33·8	80·05	174·6	364·7	745	1507
Difference per cent. ..	+33·1	+11·8	+3·1	−1·3	−3·41	−4·45

(33.) When the head is the unknown quantity, and the rest of the particulars are given, the rule becomes:—

$$\frac{\left(\frac{G}{2{\cdot}04 \times d^2} + {\cdot}0816\right)^2 - {\cdot}00665) \times \frac{L}{d}}{16{\cdot}353} = H.$$

Let us take an extreme case, in order to illustrate more fully the special adaptation of Prony's formula to very low velocities.

Say we require the head for a 10-inch pipe 4000 feet long, discharging only 20 gallons per minute: then

$$\frac{\left(\frac{20}{2\cdot 04 \times 100} + \cdot 0816\right)^2 - \cdot 00665}{16 \cdot 353} \times \frac{4000}{10} = \cdot 626 \text{ inch head.}$$

Now, by Table 3, the head comes out $\cdot 00001646 \times 1333 = \cdot 02194$ foot, or $\cdot 263$ inch only; so that in this very extreme case Prony's rule gives $\frac{\cdot 626}{\cdot 263} = 2 \cdot 38$ times the head by the rule in (5) or Table 3.

(34.) Table 29 has been calculated by the following modification of Prony's rule :—

$$\frac{(V + \cdot 0816)^2 - \cdot 00665}{196 \cdot 24} = \frac{H \times d}{L};$$

In which d = diameter of pipe in inches.
V = velocity of discharge in feet per second.
H = head of water in inches.
L = length of pipe in inches.

Table 29 has been calculated for small velocities only, because Table 3 gives results sufficiently correct for practical purposes, with higher velocities, and is more facile in application. We have added opposite each velocity in Table 29 the corresponding discharge of pipes, from 1 inch to 24 inches diameter, in order to abridge the labour as much as possible. For the use of this Table we have the following rules :—

(35.) 1st. To find the discharge, having H, L, and d given. Multiply the given head in inches by the diameter in inches, and divide by the length in inches, and find the nearest number thereto in Col. 1. Then opposite that number, and under the given diameter will be found the discharge in gallons per minute. Say, we take the case in (32) to find the discharge of a 12-inch pipe 3000 feet or 36,000 inches long, with 36 inches head. Then $\frac{H \times d}{L}$ or $\frac{36 \times 12}{36000} = \cdot 012$, the nearest number to which in

D

Col. 1 is ·01192, opposite to which, and under 12 inches diameter, is 427 gallons, the discharge sought.

2nd. To find the head, having G, L, and d given. In Table 29, under the given diameter, find the nearest number of gallons, and take from Col. 1 the number opposite to it, which number, multiplied by the length in inches, and divided by the diameter in inches, will give the required head in inches. Thus, taking the extreme case in (33) to find the head for a 10-inch pipe 4000 feet long, with 20 gallons per minute:—The nearest discharge under 10 inches diameter is 20·45 gallons, opposite which in Col. 1 is ·0001341, and from this we obtain $\dfrac{·0001341 \times 48000}{10} =$ ·643 inch head: the exact head for 20 gallons we calculated in (33) to be ·626 inch.

It should be observed that Prony's formula does not include the head due to velocity of entry (12), which for short pipes becomes important. It has been omitted in the preceding illustrations, because with such long pipes as were given in our cases it is too small to affect the result sensibly: for instance, in the last case, the head for velocity with 20 gallons per minute and a 10-inch pipe by the rule in (3) is $\left(\dfrac{20}{100 \times 13}\right)^2 =$ ·000237 foot, or $\frac{1}{352}$nd of an inch only.

(36.) "*Square and Rectangular Pipes.*"—The case of square or rectangular pipes may be assimilated to that of round ones, and the head or discharge may then be calculated by the same rules and Tables that we have given for the latter. The *velocity* of discharge, whatever may be the form of the pipe or channel, is proportional to the hydraulic radius (57) or the sectional area, divided by the circumference or perimeter: in round pipes this is always equal to one-fourth of the diameter.

Say we have a rectangular channel 3 ft. × 1·5 foot, Fig. 39; the area is 4·5 feet; the perimeter 9 feet, and the hydraulic radius $\dfrac{4·5}{9} =$ ·5 foot, which is the same as that of a round pipe ·5 × 4 = 2 feet diameter. Then to find the head for friction

EFFECT OF CORROSION OR RUST IN PIPES.

with such a channel, say 100 yards long, discharging 270 cubic feet per minute; we have a velocity of $\frac{270}{4\cdot5} = 60$ feet per minute, or 1 foot per second, which by Table 29 is equal to 1178 gallons per minute with a 24-inch pipe, and by Col. 1 of the same Table $\frac{H \times d}{L} = \cdot005928$, therefore $H = \frac{\cdot005928 \times L}{d}$ or in our case $\frac{\cdot005928 \times (100 \times 36)}{24} = \cdot889$ inch, the head required. We might have obtained the head approximately by Table 3, say for 1200 gallons $= \cdot000744 \times (100 \times 12) = \cdot8928$ inch.

We might also have calculated the head more directly by Table 30 :—Opposite ·5 the given hydraulic radius, the nearest velocity to that given, or 60 feet per minute, is 61 feet, which is under 15 inches fall per mile, or ·00852 inch per yard; hence for 100 yards the head is ·00852 × 100 = ·852 inch.

The head for velocity at entry must be added to that for friction, and may be found by Table 15 : thus, with a square-edged inlet, the head for a velocity of 1 foot per second is given by Col. C at $\frac{1}{4}$th of an inch; the total head is therefore ·889 + ·25 = 1·139 inch.

By the application of the same principles, the head, or discharge of a channel of any sectional form whatever may be determined.

(37.) "*Effect of Corrosion or Rust in Pipes.*"—The rules and Tables for calculating the discharge of pipes are adapted only to clean and even surfaces, such as are commonly met with in new cast-iron pipes. But some soft waters contain a great deal of oxygen, which rapidly decomposes iron, forming rust, which is deposited, not in an even layer, but in nodules or carbuncles.

These retard the flow, not so much by the reduction of diameter as by the alteration of the character of the surface. A notable case of this kind occurred at Torquay, where a main about 14 miles long, composed of 14,267 yards of 10-inch, 10,085 yards of 9-inch, and 170 yards of 8-inch pipe, delivered only 317 gallons per minute, with 465 feet head. We may calculate the discharge by the method explained in (13):—Assuming 1000 gallons, we have by Table 3:—

Friction of 10-inch = ·04115 × 14267 = 587·1 feet head
,,   9 ,, = ·0697 × 10085 = 702·9 ,, ,,
,,   8 ,, = ·1256 ×   170 = 21·3 ,, ,,
                                                                        1311·3 ,, total.

And from this, the discharge with the real head is $\dfrac{\sqrt{465} \times 1000}{\sqrt{1311\cdot 3}}$

or $\dfrac{21\cdot 564 \times 1000}{36\cdot 21}$ = 595 gallons. But by Prony's rule (32) the discharge comes out 616 gallons. The experimental discharge was therefore only $\dfrac{317}{616}$ = ·51 or 51 per cent. of the theoretical, or in round numbers the discharge was that due to ¼th of the head, so that ¾ths of the head was lost in undue friction. An ingenious scraper, suggested by the late Mr. Appold, and worked by the pressure of the water, was passed through the entire length of the pipes; and subsequently an improved one by W. Froude, Esq., was used with remarkable results, the discharge being increased to 564, and eventually, by repeated scraping, to 634 gallons, which is 18 gallons, or 3 per cent. more than the theoretical quantity. Errors of observation, or in the reputed sizes of the pipes, may account for the discrepancy.

Dr. Angus Smith's process, by which pipes are coated all over with a black enamel, seems to be an effective remedy against rusting; such pipes have been used with Torquay water for years without being affected. The process is very cheap, being only about 5s. per ton for medium pipes; it can be effectively applied only in the process of casting, while the pipes are new and hot. With such a smooth surface as this process produces, the discharging power must be increased in a higher ratio than the cost, so that such pipes must really be more economical than any other.

CHAPTER II.

ON FOUNTAINS, JETS, &C.

(38.) "*Height of Jets with given Heads*."—When water issues vertically from a nozzle, as at J in Fig. 5, it should theoretically attain the height of the head, and h should be equal to H; but it has been found by experiment that the height of the jet is always less than the head, a loss arising from the resistance of the air. The difference, or h', is found to increase with the absolute height of the jet, and to diminish with an increase in the diameter. There are very few reliable experiments on this subject, and the laws indicated by those we have are very intricate. The best experiments we have are given in Table 7, and from them we find that h' increases nearly in the ratio of the square of the head, so that if we draw to scale the successive heights found by experiment, as in Fig. 14, we obtain a curve which approximates to a parabola. Thus, for a ½-inch jet, as in the Figure, with 160 feet head, the jet would have attained the height B, or 160 feet, if there had been no resistance from the air; but it is found by experiment that it only reaches 80 feet as at D, therefore $h' = 80$ feet is lost. Again, with 80 feet head the jet should have reached C = 80 feet, but the experimental height is only 60 feet, and, in that case, $h' = 20$ feet. Thus with heads in the ratio of 1, 2, the loss is in the ratio 1^2, 2^2, or 1 to 4, being in fact 20 and 80 feet.

(39.) Experiment also shows, that the head being constant, h' varies nearly in inverse ratio to the diameter of the jet; for instance, we have just seen that with 80 feet head on the ½-inch jet, 20 feet head is lost. Then with a jet 1 inch diameter the loss would be about 10 feet, and the height attained 70 feet; but with a ¼-inch jet the loss would be about 40 feet, and the height attained 40 feet, &c. Thus we have the elements for calculating approximately the loss of head for any particular case, not perfectly agreeing, perhaps, with the true law, but the best

FOUNTAINS—EXPERIMENTS ON HEIGHT OF JETS.

TABLE 7.—Of EXPERIMENTS on the HEIGHT of JETS with DIFFERENT HEADS.

Diam. of Jet in Inches.	Head on the Jet in Feet.	Height of Jet in Feet.		Error.	Loss of Height by Jet in Feet.		
		Experi-ment.	Calcu-lated.		Experi-ment.	Calcu-lated.	
				feet.			
$2\frac{1}{2}$	365	284	282	−2·0	81	83	Chatsworth.
$1\frac{5}{8}$	64	61	60·1	−0·9	3	3·9	Witley Court.
,,	92	84	83·86	−0·14	8	8·14	,,
,,	115	103	102·3	−0·7	12	12·7	,,
1	445	109	136·0	+27·0	336	309	Torquay.
$\frac{3}{4}$	46	43	41·2	−1·8	3	4·8	Witley Court.
,,	69	62	59·0	−3·0	7	10·0	,,
,,	92	77	74·4	−2·6	15	17·6	,,
,,	115	93	87·5	−5·5	22	27·5	,,
,,	141	98	99·6	+1·6	43	41·4	,,
,,	162	106	107·3	+1·3	56	54·7	,,
$\frac{5}{8}$	15	14·25	14·44	+0·19	0·75	0·56	Weisbach.
,,	30	27·81	27·75	−0·06	2·19	2·25	,,
,,	45	39·42	39·94	+0·52	5·58	5·06	,,
,,	60	48·36	51·00	+2·64	11·64	9·00	,,
$\frac{3}{8}$	15	14·04	14·06	+0·02	0·96	0·94	,,
,,	30	26·44	26·25	−0·19	3·56	3·75	,,
,,	45	36·18	36·56	+0·38	8·82	8·44	,,
,,	60	42·96	45·00	+2·04	17·04	15·00	,,
,,	32	27	27·7	+0·7	5	4·3	Witley Court.
,,	46	36	37·2	+1·2	10	8·8	,,
,,	95	55	57·4	+2·4	40	37·6	,,
,,	118	63	60·0	−3·0	55	58·0	,,
$\frac{3}{16}$	28·8	19	21·9	+2·9	9·8	6·9	,,
,,	64	30	30·0	0·0	34·0	34·0	,,

approximation we can obtain: this is a subject on which more experimental information is very desirable. Table 8 gives the height of jets with different heads, and is calculated by the following rule:—

$$h' = \frac{H^2}{d} \times \cdot 0125 ;$$

In which H = the head on the jet in feet.

,, h' = the difference between the height of head and height of jet.

,, d = diameter of jet in $\frac{1}{8}$ths of an inch.

FOUNTAINS—HEIGHT OF JETS WITH DIFFERENT HEADS.

TABLE 8.—Of the HEIGHT of JETS with DIFFERENT HEADS.

\multicolumn{10}{c}{Diameter of Jet in Inches.}
$\frac{1}{5}$	$\frac{1}{4}$	$\frac{3}{8}$	$\frac{1}{2}$	$\frac{5}{8}$	$\frac{3}{4}$	1	$1\frac{1}{4}$	$1\frac{1}{2}$	$1\frac{3}{4}$	2

Height of Jet in Feet.

Head	$\frac{1}{5}$	$\frac{1}{4}$	$\frac{3}{8}$	$\frac{1}{2}$	$\frac{5}{8}$	$\frac{3}{4}$	1	$1\frac{1}{4}$	$1\frac{1}{2}$	$1\frac{3}{4}$	2
8 75	9·37	9·6	9·7	9·75	9·8	9·84	9·875	9·9	9·91	9·92
15 0	17·5	18·33	18·75	19·0	19·2	19·4	19·5	19·6	19·6	19·7
19 0	24·4	26·25	27·2	27·75	28·3	28·6	29·0	29·1	29·2	29·3
20 0	30·0	33·3	35·0	36·0	37·0	37·5	38·0	38·3	38·6	38·7
,,	34·4	39·6	42·2	44·0	45·0	46·1	47·0	47·4	47·8	48·0
,,	37·5	45·0	48·7	51·0	52·0	54·4	55·0	56·2	58·6	57·0
,,	39·0	50·0	55·0	58·0	60·0	62·4	64·0	65·0	65·6	66·0
,,	40·0	53·0	60·0	64·0	67·0	70·0	72·0	73·3	74·2	75·0
,,	..	56·0	65·0	70·0	73·0	77·0	80·0	81·6	83·0	84·0
,,	..	58·0	69	75	79	84	87	90	91	92
,,	..	60·0	75	84	90	97	102	105	107	109
,,	..	..	79	91	99	109	116	120	123	125
,,	..	..	80	96	106	120	128	133	137	140
,,	..	..	..	99	112	129	139	141	151	155
,,	..	..	..	100	116	137	150	158	166	169
,,	..	..	..	..	119	145	159	165	177	182
,,	..	..	..	..	120	150	168	180	189	195
,,	..	..	..	..	..	155	175	190	200	208
,,	..	..	..	..	..	158	182	198	210	219
,,	..	..	..	..	..	160	187	206	220	230
,,	..	..	..	..	..	..	198	222	241	255
,,	..	..	..	..	..	..	200	233	257	275

(40.) It is a result of this rule, that each particular size of jet attains its maximum height with a certain head, and that if the head is increased beyond that point, the height of jet is not increased thereby, but is actually diminished. This result is anomalous: it may be that an excessive head breaks the issuing stream into spray and causes it to meet with more resistance from the air than a jet of solid water issuing with a moderate head. Experiments with excessive heads show an enormous loss: thus a jet 1 inch diameter with 445 feet head, reached a height of about 109 feet only, as measured by a theodolite. Our rule gives the loss $h' = \dfrac{445^2}{8} \times \cdot 0125$, or $\dfrac{198025}{8} \times \cdot 0125$

= 309 feet, and hence the height of jet is 445 − 309 = 136 feet. The error of 27 feet is considerable, but perhaps not more than might be expected in such an extreme case.

(41.) "*Discharge of Jets.*"—The quantity of water discharged will vary considerably with the *form* of the nozzle. The form is also a matter of importance, as affecting the solidity of the issuing stream, and thereby the height of the jet. Fig. 15 shews the best form of nozzle, and Table 9 gives the general proportions

TABLE 9.—Of the PROPORTIONS OF NOZZLES for JETS.

A.	B.	C.	D.
in.	in.	in.	in.
$\frac{1}{4}$	·45	·6	·3
$\frac{3}{8}$	·67	·9	·45
$\frac{1}{2}$	·90	1·2	·6
$\frac{5}{8}$	1·12	1·5	·75
$\frac{3}{4}$	1·35	1·8	·9
1	1·80	2·4	1·2
$1\frac{1}{4}$	2·25	3·0	1·5
$1\frac{1}{2}$	2·70	3·6	1·8
$1\frac{3}{4}$	3·15	4·2	2·1
2	3·6	4·8	2·4
$2\frac{1}{4}$	4·0	5·4	2·7
$2\frac{1}{2}$	4·5	6·0	3·0
$2\frac{3}{4}$	4·9	6·6	3·3
3	5·4	7·2	3·6

for different sizes. The lip at E projecting beyond the mouth is intended to protect the bore from indentation by accident. The discharge by well-made nozzles of this form will be about ·943, the theoretical discharge being 1·0, and may be found direct by the following rule:—

$$G = \sqrt{H} \times d^2 \times \cdot 24;$$

In which H = the head of water on the jet in feet.
        d = the diameter in $\frac{1}{8}$ths of an inch.
        G = gallons discharged per minute.

Table 10 has been calculated by this rule.

(42.) "*Jets at the End of Long Mains.*"—When a jet is placed at the end of a pipe, or series of pipes, as is usually the case,

FOUNTAINS—DISCHARGE OF JETS. 41

TABLE 10.—Of the Discharge of Jets with Different Heads.

Head on Jet in Feet.	Diameter of Jet in Inches.														
	1/8	3/16	1/4	5/16	3/8	7/16	1/2	5/8	3/4	7/8	1	1¼	1½	1¾	2	2¼	2½

						Gallons Discharged per Minute.										
5	·537	1·21	2·15	3·36	4·83	6·58	8·59	13·4	19·3	26·3	34·4	53·7	77·0	105	137	174	215
10	·758	1·71	3·03	4·74	6·82	9·30	12·1	18·9	27·3	37·1	48·5	75·8	109	148	194	244	303
15	·929	2·09	3·72	5·81	8·36	11·4	14·8	23·2	33·4	45·5	59·4	92·9	134	182	238	301	372
20	1·07	2·41	4·29	6·70	9·66	13·0	17·2	26·8	38·6	52·6	68·6	107	154	210	275	347	429
25	1·20	2·70	4·80	7·50	10·8	14·7	19·2	30·0	43·2	58·8	76·8	120	173	235	307	389	480
30	1·31	2·95	5·25	8·21	11·8	16·1	21·0	32·8	47·3	64·4	84·1	131	189	258	336	426	525
35	1·42	3·19	5·68	8·87	12·8	17·4	22·7	35·5	51·1	69·6	90·9	142	204	278	364	460	568
40	1·52	3·41	6·07	9·48	13·6	18·6	24·7	37·9	54·6	74·3	97·1	152	218	297	388	491	609
45	1·61	3·62	6·44	10·1	14·5	20·2	25·8	40·3	58·0	80·7	103	161	232	323	412	522	644
50	1·70	3·82	6·79	10·6	15·2	21·3	27·1	42·4	61·1	84·9	109	170	244	340	434	550	680
60	1·86	4·18	7·44	11·6	16·7	22·8	29·7	46·4	66·9	91·1	119	186	267	364	476	602	744
70	2·01	4·52	8·03	12·5	18·1	24·5	32·1	50·1	72·3	98·4	129	201	289	393	514	650	803
80	2·14	4·83	8·58	13·4	19·3	26·3	34·3	53·6	77·2	105	137	215	309	421	549	695	858
90	2·27	5·12	9·10	14·2	20·5	27·9	36·4	56·9	82·0	111	145	227	328	446	583	738	910
100	2·40	5·40	9·6	15·0	21·6	29·4	38·4	60·0	86·4	117	153	240	346	471	615	778	960
110	2·52	5·66	10·1	15·7	22·6	30·9	40·3	62·9	90·6	123	161	252	362	493	644	815	1010
120	2·63	5·91	10·5	16·4	23·6	32·6	42·0	65·7	94·6	130	168	263	378	522	673	852	1052
130	2·73	6·15	10·9	17·1	24·6	33·5	43·7	68·4	98·5	134	175	273	394	536	700	886	1094
140	2·84	6·39	11·3	17·7	25·5	34·4	45·4	71·0	102	139	181	284	409	557	727	920	1136
150	2·94	6·61	11·7	18·4	26·4	36·0	47·0	73·5	106	144	188	294	423	576	752	952	1176
175	3·17	7·14	12·7	19·8	28·5	39·0	50·8	79·4	114	156	203	317	457	624	813	1029	1270
200	3·39	7·63	13·5	21·2	30·5	41·5	54·3	84·8	122	166	217	339	490	665	869	1099	1358
250	3·79	8·53	15·1	23·7	34·1	46·5	60·7	94·8	136	186	243	379	546	744	971	1230	1518
300	4·15	9·35	16·6	25·9	37·4	50·1	66·5	104	149	204	266	415	598	816	1064	1346	1663

calculation must be made of the loss of head by friction in such pipes, so as to obtain the actual head *on the jet,* for which alone the rules and Table apply. Say, for illustration, we take the case, shown by Fig. 16, of a jet 1 inch diameter, 70 feet high, at the end of a long main 6 inches, 5 inches, and 4 inches diameter, of the respective lengths given by the Figure, and that we have to calculate the head necessary. Table 8 shows that a jet 1 inch diameter, 70 feet high, requires 80 feet head; and Table 10 gives the discharge of the same jet, with 80 feet head, at 137 gallons. Then, by Table 3, we calculate the friction of the mains, and we have the following results:—

	Feet.
Head to play 1-inch jet 70 feet high $= $	80·00
Friction 6-inch main, say 140 gallons $= ·01037 \times 600 =$	6·22
,, 5 ,, ,, $= ·0258 \times 300 =$	7·74
,, 4 ,, ,, $= ·0788 \times 100 =$	7·88
Total $=$	101·84

(43.) In other cases we may have the head and diameter of pipes and nozzle given, and have to determine the discharge. This case is illustrated by Fig. 17, and in dealing with it, we must follow the course indicated in (13). Say we assume the discharge at 300 gallons; Table 10 shows that a jet $1\frac{1}{2}$ inch diameter requires about 75 feet head for that quantity. Then, by Table 3, we find the friction of the mains as follows:—

	Feet.
Head to play $1\frac{1}{2}$-inch jet, 300 gallons $=$	75·00
Friction 7-inch main, 300 gallons $= ·022 \times 800 =$	17·60
,, 6 ,, ,, $= ·0476 \times 400 =$	19·04
,, 5 ,, ,, $= ·1185 \times 80 =$	9·48
Total $=$	121·12

So that for our assumed discharge of 300 gallons we require only 121·12 feet, instead of 150, the head at disposal. Then by the rule in (13) the true discharge with 150 feet head will be $\dfrac{300 \times \sqrt{150}}{\sqrt{121·12}} = 334$ gallons. In such cases as this, where the height of a jet is involved, the discharge assumed should be pretty near the true one.

(44.) In another case we might require to find the diameter of one of the main pipes, having all the rest given. Thus, say that we have to find the diameter of the pipe P, in Fig. 18. Table 8 gives 90 feet as the head for $1\frac{1}{4}$ jet 80 feet high; and Table 10 gives 227 gallons as the discharge of the same jet with 90 feet head.

Then, $1\frac{1}{4}$ jet 80 feet high, by Table 8 .. 90·0 feet head
Friction of 6-inch main = ·028 × 400 .. $\underline{11·2}$ „
101·2 „

We have therefore $115 - 101·2 = 13·8$ feet of head left for the friction of the pipe P, or $\frac{13·8}{200} = ·069$ foot per yard; which by Table 3 is equal to a 5-inch pipe with say 230 gallons, and this is the required diameter of the pipe P.

(45.) "*Path of Fountain Jets.*"—When the discharge takes place obliquely, or out of the perpendicular, the path of the jet is a parabola, and may be conveniently described by the method shown in Fig. 23, in which we have a jet discharging upward at an angle of 45°, and with a head of 14 feet, which by Table 11 will give a velocity of 30 feet per second, or 3 feet per tenth of a second. If we mark on the line S, E a series of points A, B, C, &c., 3 feet apart, they would show the position of a particle of water at each tenth of a second if gravity did not act: but of course gravity does act simultaneously, and Table 12 gives the space fallen through each tenth of a second, which, being plotted on the perpendiculars drawn through each of the points A, B, C, &c., will give the true position of the particle of water at each tenth of a second. Thus, in $\frac{3}{10}$ths of a second it would have arrived at C, if uninfluenced by gravity, but the Table shows that in that time a body falls 1 foot $5\frac{1}{4}$ inches; therefore F is the true position at that moment, and so of the rest, as in the Figure, which gives the path for two seconds. The lower curve S, T in Fig. 23, shows the path of a jet with the same head and velocity projected *downwards* at the same angle of 45°. Fig. 19 gives the path for a horizontal projection, and also

FOUNTAINS—TABLES OF VELOCITIES.

TABLE 11.—FALLING BODIES, giving the SPACE fallen through to acquire certain VELOCITIES.

Velocity in Feet per Second.	Space.		Velocity in Feet per Second.	Space.		Velocity in Feet per Second.	Space.	
	ft.	ins.		ft.	ins.		ft.	ins.
1	0	$0\frac{3}{16}$	21	6	10	41	26	1
2	0	$0\frac{3}{4}$	22	7	6	42	27	5
3	0	$1\frac{5}{8}$	23	8	3	43	28	9
4	0	3	24	9	0	44	30	1
5	0	$4\frac{5}{8}$	25	9	9	45	31	5
6	0	$6\frac{3}{4}$	26	10	6	46	32	10
7	0	$9\frac{1}{8}$	27	11	4	47	34	4
8	1	0	28	12	3	48	36	10
9	1	$3\frac{1}{4}$	29	13	0	49	37	4
10	1	$6\frac{3}{4}$	30	14	0	50	38	11
11	1	$10\frac{1}{2}$	31	14	11	52	42	0
12	2	3	32	15	11	54	45	4
13	2	$7\frac{1}{2}$	33	16	11	56	50	0
14	3	$0\frac{3}{4}$	34	18	0	58	52	0
15	3	6	35	19	0	60	56	0
16	4	0	36	20	1	62	59	8
17	4	6	37	21	5	64	63	8
18	5	0	38	22	6	66	67	8
19	5	7	39	23	9	68	72	0
20	6	3	40	24	11	70	76	0

TABLE 12.—FALLING BODIES.

Time. Seconds.	Whole Space fallen.		Velocity acquired. Feet per Second.	Time. Seconds.	Whole Space fallen.		Velocity acquired. Feet per Second.
	ft.	ins.	ft.		ft.	ins.	ft.
$\frac{1}{10}$	0	$1\frac{9}{16}$	3·2	$1\frac{1}{10}$	19	$4\frac{3}{8}$	35·2
$\frac{2}{10}$	0	$7\frac{5}{8}$	6·4	$1\frac{2}{10}$	23	$0\frac{1}{2}$	38·4
$\frac{3}{10}$	1	$5\frac{1}{4}$	9·6	$1\frac{3}{10}$	27	$0\frac{1}{2}$	41·6
$\frac{4}{10}$	2	$6\frac{3}{4}$	12·8	$1\frac{4}{10}$	31	$4\frac{3}{8}$	44·8
$\frac{5}{10}$	4	0	16·0	$1\frac{5}{10}$	36	0	48·0
$\frac{6}{10}$	5	$9\frac{1}{8}$	19·2	$1\frac{6}{10}$	41	0	51·2
$\frac{7}{10}$	7	10	22·4	$1\frac{7}{10}$	46	$2\frac{5}{8}$	54·4
$\frac{8}{10}$	10	$2\frac{7}{8}$	25·6	$1\frac{8}{10}$	51	1	57·6
$\frac{9}{10}$	12	$11\frac{1}{2}$	28·8	$1\frac{9}{10}$	57	$9\frac{1}{8}$	60·8
1	16	0	32·0	2	64	0	64·0

illustrates another method of drawing the parabolic curve, which consists in dividing the total space fallen through J, K into the same number of equal parts as the line H, J, and drawing radial lines from the point H, as shown. The path of the jet is through the intersections of the radial lines with the perpendiculars, as in the figure: the two methods give the same result precisely.

(46.) There are some general laws governing the parabolic paths of jets which it will be well to state explicitly. Let Fig. 20 be a jet playing obliquely from a nozzle at J, and striking the horizontal plane at G.

1st. If the line of direction of the pipe or axis of the jet be prolonged, it cuts the axis of the parabola at a point C, whose distance from the base is always double the height of the parabola, or C N is equal to twice D N. This gives a useful rule for finding the proper angle of the jet pipe when the path of the jet has been determined.

2nd. If we find the focus of the parabola by the ordinary method, namely, by bisecting the radius of the base at A, drawing the line A D, and making A L perpendicular to A D, then the point L is the focus of the parabola and the distance N L is the *extra* head h necessary to play the jet horizontally, or the difference between the maximum height of the jet and the head upon it at J. Thus the total head H' may be considered as divided into two portions, namely, H, which is equal to the height of the parabola D N, and h, which is equal to the distance of the focus of the parabola from the base.

3rd. If, therefore, with the same head the jet were made to play vertically, it would (theoretically) attain the height of H', instead of H.

4th. In all cases, h bears a certain proportion to the height of the parabola (H), and to the length of its base B, and may be calculated from those particulars by the rule $h = \dfrac{(\frac{1}{4} B)^2}{H}$; thus, to play a jet 32 feet horizontally (B), and 16 feet high (H), as in Fig. 21, we shall have $h = \dfrac{8^2}{16} = 4$ feet, which, added to the height of the jet path (16 feet), gives 20 feet for the total head on the jet.

5th. The horizontal distance from the nozzle at J to the point on the plane at G, where the jet strikes it, may be calculated when the total head H' and the height of the parabola H are given; for obviously $H' - H = h$, and knowing h, we may find B by the rule $\sqrt{h \times H} \times 4 = B$. Thus, in Fig. 21, we have $H' = 20$, and $H = 16$; therefore, $h = 20 - 16 = 4$, and then $\sqrt{4 \times 16} \times 4 = 32$ feet.

6th. When the jet issues horizontally, as in Fig. 25, its path is half a parabola, following the same laws as before, namely, $h = F$, also $h = \dfrac{(\frac{1}{2} P)^2}{H}$, and $\sqrt{h \times H} \times 2 = P$, &c.

(47.) In some cases, the two half parabolas are unequal, as in Fig. 24, where we have a jet 20 feet high at its maximum, delivering at $N = 15$ feet high, and 24 feet distant horizontally from the nozzle at J, and we require to find $h =$ the extra head, and to describe the path of the jet. Here we have first to find the position of the centre line dividing the semi-parabolas, and to do this we have $\dfrac{D \times \sqrt{H}}{\sqrt{H} + \sqrt{H''}} = R$, which in our case becomes $\dfrac{24 \times 4\cdot472}{4\cdot472 + 2\cdot236} = 16$ feet. Then the focus of the two semi-parabolas may be found as before, and it will be found that F and F' are equal. Thus, in our case $F = \dfrac{\left(\frac{16}{2}\right)^2}{20} = 3\cdot2$ feet, and $F' = \dfrac{\left(\frac{8}{2}\right)^2}{5} = 3\cdot2$ feet also. F being equal to h, we thus find h to be 3·2 feet, and the total head at J will therefore be $20 + 3\cdot2 = 23\cdot2$ feet (H'). If we reverse the direction of the jet, placing the nozzle at N, instead of at J, then, with a head of $5 + 3\cdot2 = 8\cdot2$ feet, the path of the jet would be the same as before.

(48.) We have followed throughout the investigation of the paths of oblique jets, the theoretical law that the height of the jet is equal to the head, and we have done this to avoid complicating the matter unnecessarily; but obviously, we must apply to oblique jets the correction we found necessary for perpendicular ones. Thus, if we had a jet ½-inch diameter, with 80 feet head, Table 8 shows that the height attained vertically would be only 60 feet, and if this jet played obliquely, its path should be calculated for the latter height, but the quantity of water expended, and the value of h must be calculated for 80 feet.

Oblique jets of great height and range, deviate considerably from the true parabolic path assigned by the rules; the curve becomes in such cases like A, D, E in Fig. 22, the true parabolic path being A, B, C. But for moderate heights and ranges, such as usually occur in practice, the deviation is not considerable.

(49.) "*Ornamental Jets.*"—There are many kinds of ornamental jets which may be used with pleasing effect in *very sheltered* situations, especially in the interior of conservatories, &c. One of these, called the "Convolvulus," from the form of its display, is shown in half-size section by Fig. 26. The pressure of a very small head of water (2 or 3 feet) raises the valve B, and allows a thin sheet of water to escape, forming a sheet jet of the form given in Fig. 27, and (with the size given by Fig. 26) about 3 feet diameter, with an expenditure of about 6 gallons per minute.

Fig. 28 is a half-size section of the "Dome" or "Globe" jet, which produces a display of the form shown by Fig. 29, with a head of about 2 feet, the globe being about 14 inches diameter, and the expenditure about 3 gallons per minute. With a greater head, say 3 or 4 feet, the display has the form of an umbrella about 21 inches diameter, expending about 4 gallons per minute.

The "Basket and Ball" jet is another pleasing variety; the basket is of fancy wire-work, large enough to catch the ball when it escapes from the jet of water, and formed so as to return it back to its place. The ball is formed of light wood (lime-tree is the best), painted or gilded, and well varnished.

There should be a certain proportion between the size of the ball and the diameter of the jet. As an approximation we may give the following rule:—

$$\sqrt[3]{d^2 \times 1\cdot 3} = D;$$

In which d = the diameter of the jet in $\frac{1}{8}$ths of an inch.
D = the diameter of the ball in inches.

Table 13 has been calculated by this rule; it gives the proportions up to 1-inch jets, but the $\frac{3}{4}$-inch jet, with $3\frac{1}{2}$-inch ball is usually the maximum size in practice.

TABLE 13.—For BALL JETS.

Diameter of Jet.		Diameter of Ball.	
$\frac{1}{8}$-inch	=	$1\frac{1}{8}$-inch	
$\frac{1}{4}$	,,	=	$1\frac{3}{4}$	,,
$\frac{3}{8}$	,,	=	$2\frac{1}{4}$	,,
$\frac{1}{2}$	,,	=	$2\frac{3}{4}$	,,
$\frac{5}{8}$	,,	=	$3\frac{1}{8}$	,,
$\frac{3}{4}$	,,	=	$3\frac{1}{2}$	,,
$\frac{7}{8}$	,,	=	4	,,
1	,,	=	$4\frac{3}{8}$	,,

CHAPTER III.

ON CANALS, CULVERTS, AND WATER-COURSES.

(50.) "*Open Water-courses.*"—The discharge of open water-courses may be found experimentally by observing the velocity of the current and measuring the cross sectional area of the stream. But to do this correctly we require the *mean* velocity throughout the section, which is not given by observation. The velocity varies, being a maximum at the surface and where the channel is deepest, which is usually near the centre of the width, diminishing from thence to the banks on either side, and to the bottom, where it is a minimum.

The best experiments we have, give the mean velocity

CANALS—HEAD DUE TO **VELOCITY**. 49

throughout the **section** at 84 per **cent. of** the maximum central surface **velocity, which is** usually the velocity observed, being easily **obtained** by **a** float on the surface of the stream (68). **Table 14** gives the mean velocity corresponding to observed **maximum** velocities; thus, if a **channel** whose area is 24 square **feet, has** by observation a central **surface** velocity of 35 feet per minute, the mean velocity **by the Table is** 29·4 feet, and the discharge will be 29·4 × 24 = **705·6** cubic feet, or 705·6 × **6·23** = 4396 gallons per minute.

TABLE 14.—For OPEN **CHANNELS,** CANALS, and **RIVERS,** giving the MEAN VELOCITY **throughout** the SECTION, corresponding to observed CENTRAL SURFACE VELOCITIES.

Surface Velocity.	Mean Velocity.	Surface Velocity.	Mean Velocity.	Surface Velocity.	Mean Velocity.	Surface Velocity.	Mean Velocity.
1	·84	26	21·84	51	42·84	76	63·84
2	1·68	27	22·68	52	43·68	77	64·68
3	2·52	28	23·52	53	44·52	78	65·52
4	3·36	29	24·36	54	45·36	79	66·36
5	4·2	30	25·2	55	46·20	80	67·2
6	5·04	31	26·06	56	47·04	81	**68·04**
7	5·88	32	26·88	57	47·88	82	**68·88**
8	6·72	33	27·72	58	48·72	83	**69·72**
9	7·56	34	28·56	59	49·56	84	**70·56**
10	8·4	35	29·4	60	50·4	85	**71·40**
11	9·24	36	30·24	61	51·24	86	**72·24**
12	10·08	37	31·08	62	52·12	87	73·08
13	10·92	38	31·92	63	52·92	88	73·92
14	11·76	39	32·76	64	53·76	89	74·76
15	12·60	40	33·6	65	54·6	90	75·6
16	13·44	41	34·44	66	55·44	91	76·44
17	14·28	42	35·28	67	56·28	92	77·28
18	15·12	43	36·12	68	57·12	93	78·12
19	15·96	44	36·96	69	57·96	94	78·96
20	16·8	45	37·8	70	58·8	95	79·80
21	17·64	46	38·64	71	59·68	96	**80·64**
22	18·48	47	39·48	72	60·48	97	**81·48**
23	19·32	48	40·32	73	61·32	98	**82·32**
24	20·16	49	41·16	74	62·16	99	**83·16**
25	21·0	50	42·0	75	63·00	100	**84·00**

(51.) "*Head due to Velocity in Open Channels.*"—When a stream leaves the still water of a lake or reservoir, as in Fig. 30,

E

at a given velocity, there will be a certain loss of head to generate that velocity, that is to say, the stream at F must be lower than the still water at E in order to create the velocity required at G. In a case like the Figure, the bottom of the channel at F being at the same level as the bottom of the reservoir at E, and with a well-rounded entrance, the velocity would be ·96 of that due to gravity, and the same co-efficient would apply to the waterway of a sluice-gate, like Fig. 31, if the gate is drawn up completely out of the water, and to the openings of a bridge with pointed piers, as at Fig. 32, the conditions being evidently similar in all the three cases. With similar conditions, but with square corners at the sides of the inlet opening, as in Fig. 34, the bottom of the channel being still at the same level as that of the reservoir, the velocity at G would be ·86 of that due to gravity, or to the difference of level between E and F, and the same co-efficient applies to the openings of a bridge with square piers as in Fig. 33.

With an opening in a sluice-gate of small thickness, as at Fig. 35, the head of water being above the lower edge of the gate, the velocity is only ·635 of that due to gravity, a contraction (2) occurring on all the four sides of the aperture. If the gate be fully drawn up, the opening becomes a weir, as at Fig. 36, then contraction occurs on three sides only, and the co-efficient rises to ·667. These co-efficients are given by Eytelwein, and Table 15 gives the velocities for different heads calculated by them.

(52.) "*Head to overcome Friction of Channel.*"—When the channel is a long one, there is not only a loss of head due to the velocity, but also a further loss by friction against the sides and bottom. Where the channel is of equal cross-sectional area from end to end, the loss of head increases uniformly from end to end, and the surface of water has a certain slope or fall per yard, or per mile. Fig. 37 shows the section of a water-course in which the fall from the still water in the reservoir at A to the point B is due to the velocity at B, and this would be the same whatever the length of the channel; its amount varies with the form of the entrance as explained in (51). From B to

CANALS—HEAD TO OVERCOME FRICTION.

C there will be a regular slope when the area of the channel is uniform, and the fall C D is due to friction for the length B C.

TABLE 15.—Of the VELOCITIES in FEET per SECOND, due to given HEADS.

Head in Inches.	A. Coef. 1·0.	B. Coef. ·96.	C. Coef. ·86.	D. Coef. ·635.	Head in Inches.	A. Coef. 1·0.	B. Coef. ·96.	C. Coef. ·86.	D. Coef. ·835.
$\frac{1}{32}$	·29	·2784	·2494	·18415	1	2·317	2·2224	1·9930	1·4713
$\frac{1}{16}$	·41	·3936	·3524	·2603	$1\frac{1}{4}$	2·590	2·4864	2·2270	1·6446
$\frac{1}{8}$	·58	·5568	·4988	·3683	$1\frac{1}{2}$	2·837	2·7235	2·4398	1·8015
$\frac{1}{4}$	·82	·7872	·7052	·5207	$1\frac{3}{4}$	3·065	2·9424	2·6360	1·9463
$\frac{5}{16}$	1·0	·9600	·8600	·6350	2	3·276	3·145	2·8174	2·0803
$\frac{1}{2}$	1·158	1·1117	·9959	·7353	$2\frac{1}{4}$	3·475	3·336	2·9885	2·2066
$\frac{5}{8}$	1·295	1·2432	1·1140	·8223	$2\frac{1}{2}$	3·663	3·516	3·1502	2·3260
$\frac{3}{4}$	1·418	1·3613	1·2195	·9004	$2\frac{3}{4}$	3·842	3·688	3·3041	2·4397
$\frac{7}{8}$	1·532	1·4707	1·3175	·9728	3	4·012	3·851	3·4503	2·5476
$\frac{15}{16}$	1·638	1·5725	1·4087	1·0401	$3\frac{1}{4}$	4·176	4·009	3·5914	2·6517
$\frac{16}{16}$	1·737	1·6675	1·4938	1·1030	$3\frac{1}{2}$	4·334	4·161	3·7272	2·7521
	1·831	1·7577	1·5747	1·1627	$3\frac{3}{4}$	4·486	4·306	3·8580	2·8486
$1\frac{1}{8}$	1·921	1·8442	1·652	1·2198	4	4·633	4·448	3·9844	2·9420
$\frac{3}{4}$	2·006	1·9258	1·725	1·2738	$4\frac{1}{2}$	4·914	4·717	4·2260	3·1204
$1\frac{3}{8}$	2·088	2·0045	1·796	1·3259	5	5·180	4·973	4·455	3·2893
$\frac{7}{8}$	2·167	2·0803	1·863	1·376	$5\frac{1}{2}$	5·433	5·216	4·672	3·450
$1\frac{7}{8}$	2·243	2·1533	1·929	1·424	6	5·675	5·448	4·881	3·6036

(53.) This **fall may be calculated by** the following rule:—

$$F = \frac{\left(\frac{C}{A}\right)^2 \times L \times P}{874520 \times A};$$

In which L = length of the channel in yards.
        A = cross-sectional area of the stream in square feet.
        P = the perimeter, or wetted border in feet.
        F = the fall, or difference of level at the two ends of the channel in inches.
        C = cubic feet discharged per minute.

Thus, in the case shown by Fig. 38, A being $6 \times 2·5 = 15$ square feet, $P = 2·5 + 6 + 2·5 = 11$ feet, say that with such a channel 1760 yards, or one mile long, we require the fall to discharge 1105 cubic feet per minute: then by the rule we have in our case $\dfrac{\left(\dfrac{1105}{15}\right)^2 \times 1760 \times 11}{874520 \times 15}$ = 8 inches fall.

(54.) To this has to be added the head for the velocity at entry, or A B in Fig. 37. The *mean* velocity being $\dfrac{1105}{15}$ = 73·66 feet, the maximum (50) will be $\dfrac{73·66}{·84}$ = 87·7 feet per minute, or 1·46 foot per second, the head for which, with square corners, is given by Col. C of Table 15 at about ½-inch. Then for a channel one mile long, the total head will be $8 + \frac{1}{2} = 8\frac{1}{2}$ inches; for ⅛th of a mile, or 220 yards, $1 + \frac{1}{2} = 1\frac{1}{2}$ inch, and for 110 yards, $\frac{1}{2} + \frac{1}{2} = 1$ inch. In the last case the head for velocity is equal to the head for friction.

(55.) When the fall is given, and the discharge has to be calculated the rule becomes :—

$$C = \left(\dfrac{874520 \times F \times A}{L \times P}\right)^{\frac{1}{2}} \times A.$$

Thus, with the same channel as before, 1760 yards long, and a fall of 12 inches, the discharge would be $\left(\dfrac{874520 \times 12 \times 15}{1760 \times 11}\right)^{\frac{1}{2}}$ × 15 = 1353 cubic feet per minute. We have omitted in this case to allow for the head due to velocity, and where the channel is a long one, the omission will not cause a serious error; with short channels, however, it must not be neglected.

(56.) When, with a given total head, we have to calculate the discharge by a channel so short that the head for velocity has to be considered as well as that due to friction, the question does not admit of a direct solution, because we cannot tell beforehand in what proportions the head at disposal has to be divided between the two. The best course in that case is to assume a discharge, and calculate, as in (53) and (54), the head for friction and the head for velocity with that discharge. Then

CANALS—DISCHARGE WITH GIVEN HEADS.

applying the law (27) that the discharges are directly proportional to the square roots of the respective heads, we may obtain the true discharge with the given head. Thus say that with the channel (Fig. 38) 50 yards long, the *total* head at disposal was 2 inches, and that we have to calculate the discharge. Say we assume it at 1000 cubic feet; then the head for friction would be

$$\frac{\left(\frac{1000}{15}\right)^2 \times 50 \times 11}{874520 \times 15} = \cdot 186 \text{ inch.}$$

The mean velocity being $\frac{1000}{15} = 66 \cdot 7$, the maximum will be $\frac{66 \cdot 7}{\cdot 84} = 79 \cdot 3$ feet per minute, or $1 \cdot 32$ foot per second, the head for which by Col. C in Table 15 is about $\frac{7}{16}$ or $\cdot 437$ inch; the total head for 1000 cubic feet is, therefore, $\cdot 186 + \cdot 437 = \cdot 623$ inch: hence the discharge with 2 inches head would be $\frac{1000 \times \sqrt{2}}{\sqrt{\cdot 623}}$ or $\frac{1000 \times 1 \cdot 414}{\cdot 7893} = 1791$ cubic feet per minute.

Checking this result by the rule in (53) &c., we find that the head for friction is about $\cdot 6$ inch, and for velocity $1 \cdot 4$ inch. If in this case the head for velocity had been neglected, and the full head of 2 inches had been allowed for friction alone, the discharge would have come out $\left(\frac{874520 \times 2 \times 15}{50 \times 11}\right)^{\frac{1}{2}} \times 15 = 3276$ cubic feet, instead of 1791, the true discharge. This will serve to show the importance of considering the head for velocity with short channels.

(57.) Table 30 has been calculated by the following modification of the rule :—

$$V = \left(F \times R \times 497\right)^{\frac{1}{2}}$$

In which V = mean velocity in feet per minute.
        F = the fall in inches per mile.
        R = hydraulic radius, or area in square feet, divided by border in feet.

54 RIVER-CHANNELS OF IRREGULAR CROSS-SECTIONS.

The use of this Table may be illustrated by the following examples :—Say we calculate by it the discharge of the channel (Fig. 38) with a fall of 12 inches per mile as in (55). The hydraulic radius in our case is $\frac{15}{11} = 1\cdot 363$ foot, the nearest radii to which in the Table we find to be $1\cdot 3$ and $1\cdot 4$, and the corresponding velocities under the fall of 12 inches per mile are $88\cdot 1$ and $91\cdot 4$ respectively; interpolating between those numbers for our radius $1\cdot 363$ we find the mean velocity to be about $90\cdot 2$ feet, and the discharge $90\cdot 2 \times 15 = 1353$ cubic feet per minute.

Again, to find the fall with the same channel 800 yards long for 1230 cubic feet per minute :—The mean velocity being $\frac{1230}{15}$ = 82 feet per minute, we look between $1\cdot 3$ and $1\cdot 4$ radii in the Table for that velocity, and we find it to be under the fall of 10 inches per mile, or $\cdot 00568$ inch per yard; hence the fall in our case is about $\cdot 00568 \times 800 = 4\cdot 54$ inches for friction alone, or C D in Fig. 37.

(58.) Take another case, shown by Fig. 40, of an open cutting with sloping banks, and say that we require the discharge with a fall of 8 inches per mile. The area being $\frac{30 + 20}{2} \times 2\cdot 5 = 62\cdot 5$ square feet, and the border $5\cdot 6 + 20 + 5\cdot 6 = 31\cdot 2$ feet, the hydraulic radius is $\frac{62\cdot 5}{31\cdot 2} = 2$, which, by Table 30, with a fall of 8 inches per mile will have a velocity of $89\cdot 2$ feet, and a discharge of $89\cdot 2 \times 62\cdot 5 = 5575$ cubic feet per minute.

(59.) "*River Channels of irregular Cross-section.*"—The application of the rules to the discharge of a stream of the natural irregular form of section may be illustrated by Fig. 41. We found in (68) that the area was $27\cdot 74$ square feet; taking say 2 feet in the compasses, and stepping along the border, we find it to measure about $24\cdot 5$ feet, the hydraulic radius is, therefore, $\frac{27\cdot 74}{24\cdot 5} = 1\cdot 132$ foot. Then, with a fall of say 10 inches per mile, Table 30 gives, opposite the radius of 1·1 (which is the nearest to the one we require), the mean velocity of 73·9 feet per minute; hence the discharge is 73·9 × 27·74 = 2050 cubic feet per minute. With a very short channel, allowance should be made for velocity at entry, as explained in (56).

Table 30 may also be applied to the calculation of the discharge, &c., of common pipes running full, or to those of a square or other section, for an illustration of which see (36), also to culverts, &c., partially filled, see (62).

(60.) "*Openings of Bridges, &c.*"—The head lost by a stream in passing through a bridge is principally that due to velocity alone, the length of the channel being in most cases so short as to have little influence on the discharge. The head for velocity may be calculated by Table 15: say we take the case (58) of the stream (Fig. 40) discharging 5575 cubic feet per minute, and passing through an opening at a bridge, say 8 feet wide and 3 feet deep. The area being 8 × 3 = 24 square feet, the velocity will be $\frac{5575}{24 \times 60}$ = 3·87 feet per second, which, with pointed piers (Fig. 32) will require by Col. B of Table 15, 3 inches head (A, B in Fig. 37). But, the stream approaches the bridge with a mean velocity of 89·2 feet, or a maximum (50) of $\frac{89·2}{·84}$ = 106 feet per minute, or 1·77 foot per second, the head due to which by the same Table is $\frac{5}{8}$ inch. The head at the bridge is, therefore, reduced to 3 − $\frac{5}{8}$ = 2$\frac{3}{8}$ inches; with square piers (Fig. 33), the head by Col. C is 3$\frac{3}{4}$ inches, or at the bridge 3$\frac{3}{4}$ − $\frac{5}{8}$ = 3$\frac{1}{8}$ inches.

(61.) "*Submerged Openings.*"—The velocity of discharge through a submerged opening A (Fig. 43) is governed by the *difference* of the level of water at the two sides of it, or by H, and is not affected by the depth below the surface at which it is placed. Table 15 will give the velocity with small heads: thus an aperture 2 feet × 1·5 foot = 3 square feet area, and with H = 5 inches, would, by Col. D of Table 15, discharge 3·2893 × 3 = 9·87 cubic feet per second.

TABLE 16.—Of the Proportions and Discharging Power of Oval Culverts.

Total Height.	Width at the Top.	Radius of the Top.	Radius of the Bottom.	Radius of the Sides.	⅘ths full of Water; to the line A in Fig. 44. Depth of Water.	⅘ths full of Water; to the line A in Fig. 44. Area in Square Feet.	⅘ths full of Water; to the line A in Fig. 44. Hydraulic Radius in Feet.	⅔rds full of Water; to the line B in Fig. 44. Depth of Water.	⅔rds full of Water; to the line B in Fig. 44. Area in Square Feet.	⅔rds full of Water; to the line B in Fig. 44. Hydraulic Radius in Feet.
ft. in.	ft. in.	ft. in.	ft. in.	ft. in.	ft. in.			ft. in.		
2 0	1 4	0 8	0 4	2 0	1 8	1·732	·442	1 4	1·303	·367
3 0	2 0	1 0	0 6	3 0	2 6	3·896	·663	2 0	2·932	·550
4 0	2 8	1 4	0 8	4 0	3 4	6·928	·884	2 8	5·213	·733
5 0	3 4	1 8	0 10	5 0	4 2	10·82	1·105	3 4	8·145	·917
6 0	4 0	2 0	1 0	6 0	5 0	15·58	1·326	4 0	11·73	1·101
7 0	4 8	2 4	1 2	7 0	5 10	21·22	1·547	4 8	15·96	1·283
8 0	5 4	2 8	1 4	8 0	6 8	27·71	1·768	5 4	20·85	1·467
9 0	6 0	3 0	1 6	9 0	7 6	35·07	1·989	6 0	26·40	1·647
10 0	6 8	3 4	1 8	10 0	8 4	43·30	2·210	6 8	32·60	1·830

OVAL CULVERTS—HEAD FOR VERY LOW VELOCITIES.

(62.) "*Discharge by Egg-shaped Culverts.*"—The discharge of culverts of the common oval or other forms may be calculated by the preceding rules, or by Table 30. The proportions of culverts are arbitrary. Fig. 44 shows a good form, and Table 16 gives the general sizes, areas, &c., when filled to two different depths, so as to adapt the Table to the varying requirements of practice. Say we take the case of a 5-feet culvert ⅝ths full of water or 4 feet 2 inches deep, with a fall of 10 inches per mile, then, by Table 16, the hydraulic radius is 1·105, and the area of waterway 10·82 feet; by Table 30 we find that with 1·1 hydraulic radius, and a fall of 10 inches per mile, the mean velocity is 73·9 feet, and the discharge 73·9 × 10·82 = 800 cubic feet per minute.

(63.) With very short culverts, allowance must be made for the velocity at entry by Table 15, &c.; thus, in the case just given, if the culvert had been only 45 yards long, the fall due to friction alone would have been, by Table 30, equal to ·00568 × 45 = ·255 or ¼ inch; the mean velocity is $\frac{73·9}{60}$ = 1·23 and the maximum $\frac{1·23}{·84}$ = 1·46 foot per second, the head due to which by Col. C of Table 15 is about ½ inch. The total head is therefore, ¼ + ½ = ¾ of an inch. To calculate with precision the discharge of short culverts, with a given fall, the method explained in (56) should be followed.

(64.) "*Head for very Low Velocities.*"—In ordinary cases Table 30 gives results sufficiently correct for practical purposes with great facility, but with very small velocities experiment has shown that the head is considerably greater than that Table would give. In such cases the more laborious and refined formulæ of Prony, Saint Venant, and Eytelwein give more correct results. A comparison of these three rules with 96 experiments on the discharge of rivers shows that Eytelwein's rule agrees best with 38 experiments, Saint Venant's with 32, and Prony's with 26. The following is a modification of Eytelwein's rule:—

58 CANALS—SPECIAL RULES FOR LOW VELOCITIES.

$$C = \left(\frac{896400 \times F \times A}{L \times P} + 42 \cdot 8\right)^{\frac{1}{2}} - 6 \cdot 534\right) \times A;$$

In which **L** = length of the channel in yards.
," **A** = cross-sectional area of the stream in square feet.
," **P** = the perimeter, or border of the channel in feet.
," **F** = the fall, or difference of level at the two ends of the channel in inches.
," **C** = cubic feet discharged per minute.

(65.) Thus, say that we require the discharge by the channel, Fig. 40, 1 mile long, with a fall of 1 inch only, then L = 1760, A = 62·5, P = 31·2, as in (58), and F = 1, and the discharge will be $\left(\frac{896400 \times 1 \times 62 \cdot 5}{1760 \times 31 \cdot 2} + 42 \cdot 8\right)^{\frac{1}{2}} - 6 \cdot 534) \times 62 \cdot 5 = 1629 \cdot 3$ cubic feet per minute. We may compare this result with that given by the rule in (55), by which the discharge comes out $\left(\frac{874520 \times 1 \times 62 \cdot 5}{1760 \times 31 \cdot 2}\right)^{\frac{1}{2}} \times 62 \cdot 5 = 1972$ cubic feet per minute = $\frac{1972}{1629} = 1 \cdot 21$, or 21 per cent. difference. But with an increased head, the difference becomes less, and is reduced practically to nothing with large heads, as shown by Table 17.

TABLE 17.—Of the DISCHARGE of an OPEN CHANNEL, Fig. 40, calculated by DIFFERENT RULES.

Fall in Inches per Mile.	Calculated Discharge.		Difference per Cent.	By Table 30.		
	By Rule in (64).	By Rule in (55).		Velocity.	Area.	Discharge.
1	1629	1972	21·0	31·5 ×	62·5 =	1969
2	2444	2788	14·1	44·6	,,	2788
3	3073	3416	11·1	54·6	,,	3413
4	3556	3943	10·9	63·0	,,	3938
5	4074	4409	8·2	70·5	,,	4406
6	4499	4830	7·3	77·2	,,	4825
8	5253	5577	6·2	89·2	,,	5575
10	5918	6235	5·3	99·7	,,	6231
12	6519	6834	4·9	109·2	,,	6825
24	9380	9649	3·0	154·4	,,	9650
36	11576	11831	2·2	189·1	,,	11819

This shows that in all cases where extreme accuracy is desired, the rule in (64) should be used; but that where the fall exceeds 8 or 10 inches per mile, Table 30 gives results sufficiently correct for most practical purposes.

(66.) When the discharge is given, to determine the fall, the rule becomes

$$F = \frac{\left(\left(\frac{C}{A} + 6 \cdot 534\right)^2 - 42 \cdot 8\right) \times L \times P}{896400 \times A}.$$

Thus the fall for friction with the same channel, Fig. 40, 2000 yards long to deliver 3000 cubic feet per minute would be

$$\frac{\left(\frac{3000}{62 \cdot 5} + 6 \cdot 534\right)^2 - 42 \cdot 8) \times 2000 \times 31 \cdot 2}{896400 \times 62 \cdot 5} = 3 \cdot 26, \text{ or } 3\tfrac{1}{4} \text{ inches.}$$

Adding the head due to velocity at entry (51), the mean velocity is $\frac{3000}{62 \cdot 5} = 48$, and the maximum $\frac{48}{\cdot 84} = 57$ feet per minute, or ·95 foot per second, the head for which by Col. C of Table 15 is about $\tfrac{1}{4}$ inch; the total head is therefore $3\tfrac{1}{4} + \tfrac{1}{4} = 3\tfrac{1}{2}$ inches.

(67.) Table 18 has been calculated by the following modification of Eytelwein's rule:—

$$\frac{(V + \cdot 1089)^2 - \cdot 0118858}{8975} = R \cdot S.$$

In which V = the mean velocity over the whole area in feet per second.

R = the hydraulic radius in feet, or $\frac{\text{area in square feet}}{\text{border in feet}}$.

S = the slope, or $\frac{\text{fall in inches}}{\text{length in inches}}$.

By this Table approximately correct results may be obtained with less labour than by the rules.

1st. To find the Velocity.—Multiply the area of the channel in square feet by the fall in inches, and divide the product by the border in feet multiplied by the length of the channel in inches: find the nearest number thereto in Col. B of Table 18, and opposite to that number in Col. A is the required velocity. Thus for the case in (65) we have $\dfrac{62 \cdot 5 \times 1}{31 \cdot 2 \times (1760 \times 36)} = \cdot 0000316$, the nearest number to which is $\cdot 00003043$ opposite $\cdot 425$ foot per second. By interpolation we may obtain a nearer approximation; for, as R. S varies nearly as V^2, we have $\left(\dfrac{\cdot 425^2 \times \cdot 0000316}{\cdot 00003043}\right)^{\frac{1}{2}}$ or $\left(\dfrac{\cdot 180625 \times \cdot 316}{\cdot 3043}\right)^{\frac{1}{2}} = \cdot 4331$ foot per second, hence the discharge comes out $\cdot 4331 \times 60 \times 62 \cdot 5 = 1624$ cubic feet per minute, or practically the same as by the rule (65).

TABLE 18.—For the DISCHARGE of CANALS, RIVERS, &c., by EYTELWEIN'S RULE.

Mean Velocity in Feet per Second.	R. S.	Mean Velocity in Feet per Second.	R. S.
·025	·0000006734	·6	·00005466
·05	·000001489	·65	·00006284
·075	·00000244	·7	·00007158
·1	·000003538	·75	·00008087
·125	·000004771	·8	·00009072
·15	·000006144	·85	·00010112
·175	·000007656	·9	·0001121
·2	·000009307	·95	·0001236
·225	·0000111	1·0	·0001357
·25	·00001303	1·1	·00016146
·275	·00001510	1·2	·0001895
·3	·00001730	1·3	·00021984
·325	·00001966	1·4	·0002524
·35	·00002214	1·5	·00028703
·375	·00002477	1·6	·00032402
·4	·00002753	1·7	·0003632
·425	·00003043	1·8	·0004047
·45	·000033484	1·9	·000448
·475	·00003666	2·0	·0004943
·5	·00003998	2·5	·000757
·55	·00004705	3·0	·001075
A	B	A	B

2nd. To find the Fall.—Divide the given discharge by the given area, and by 60, which will give the mean velocity in feet per second; find the nearest number to that in Col. A, which, multiplied by the border in feet and by the length of the channel in inches, and divided by the area in square feet will give the fall in inches. Thus, for the case in (66) we have $\frac{3000}{62\cdot 5}$ = 48 feet per minute, or $\frac{48}{60}$ = ·8 foot per second, the tabular number for which is ·00009072; then

$$\frac{\cdot 00009072 \times 31\cdot 2 \times (2000 \times 36)}{62\cdot 5} = 3\cdot 26 \text{ inches fall,}$$

as before.

68. "*Case of a Mill-stream.*"—As an example of the practical application of the rules, we will take a case in which it is desired to utilize a stream of water for driving a corn-mill. Say we have a stream 1500 yards long, with a total fall of 6 ft. 6 in. from the tail of the preceding mill. We have first to ascertain the quantity of water at disposal: selecting a spot where the current appears to be tolerably uniform for some 100 feet, and a season when the quantity is an average one according to local authorities, say we take it at a point 24 feet wide as in Fig. 41. We have then to obtain the area of the stream, and to do that, may divide the width into eight equal spaces of 3 feet each, as in the Figure, which may be done conveniently by stretching a tape across the stream: then we measure the depths midway between those divisions or at 1·5 foot, 4·5, 7·5 feet, &c., &c., using a measuring rod with a flat board about 7 or 8 inches square at the end of it, to prevent penetrating the soft bottom; and thus we obtain the series of measurements given in the figure, the mean of which we find to be 1·156 foot, the area is therefore 1·156 × 24 = 27·74 square feet. To find the velocity, two lines may be stretched across the stream near the surface, and say a "chain" or 66 feet apart, and a float being placed a few yards above the highest one, and in the centre of the width, or rather where the velocity is observed to be greatest, the exact time in passing from line to line is carefully noted. This float should be a small piece of *thin* wood, say only ¼-inch thick, so

as to be almost wholly immersed, and thus expose little surface to the action of the wind. Say that the float travels the 66 feet in 20 seconds, in one minute therefore it would be $\frac{66 \times 60}{20}$ = 198 feet. This being the maximum velocity, the mean (50) over the whole area would be $198 \times \cdot 84 = 166$ feet per minute, hence the discharge is $166 \times 27 \cdot 74 = 4600$ cubic feet per minute.

(69.) The total fall is 6 feet 6 inches; allowing 6 inches for the fall of the stream itself, the net fall at the wheel will be 6 feet; a cubic foot of water weighing $62 \cdot 3$ lbs.; the horse-power being 33,000 foot-pounds; and allowing that a breast-wheel yields 50 per cent., or $\cdot 5$ of the gross power of the water, we have $\frac{4600 \times 62 \cdot 3 \times 6 \times \cdot 5}{33000} = 26$ horse-power. A pair of 4-feet stones, grinding 4 bushels of corn per hour, requires about 4 horse-power, and a dressing-machine about 6 horse; if we allow four pairs of stones, we should require $16 + 6 = 22$ horse-power, leaving 4 horse-power for the mill-gearing and small machines, &c. The diameter of the water-wheel may be about $2 \cdot 5$ times the fall, say 15 feet, and the speed of its circumference being 4 feet per second, or 240 feet per minute, and the depth of the bucket $1 \cdot 5$ foot, the width of the wheel would be $\frac{4600}{240 \times 1 \cdot 5} = 12 \cdot 8$, say 13 feet. With other kinds of water-wheel the duty would be different: a good overshot wheel would give from 70 to 80 per cent., a breast-wheel from 45 to 60, and an undershot, in which the water acts only by its impulse, from 27 to 30 per cent.

(70.) The channel must now be altered, so as to deliver 4600 cubic feet per minute, with a fall of 6 inches in 1500 yards, or $\frac{1760 \times 6}{1500} = 7$ inches per mile. When altered to the form A, B, C, D, the area will be $\frac{24 + 12}{2} \times 3 = 54$ square feet, the *mean* velocity to discharge 4600 cubic feet will be $\frac{4600}{54} = 85 \cdot 2$

feet per minute; the border is $6\cdot 7 + 12 + 6\cdot 7 = 25\cdot 4$ feet, and the hydraulic radius $\frac{54}{25\cdot 4} = 2\cdot 126$ feet. Then by Table 30 between 2 and $2\cdot 2$ radii, the velocity $85\cdot 2$ feet is found to be under the fall of 7 inches per mile, the fall we allowed. It should be observed that it is imperative that the slope shall be *uniform* from end to end, at least where the area of the channel is uniform.

CHAPTER IV.

ON WEIRS, OVERFLOW-PIPES, &c.

(71.) "*Weirs.*"—Fig. 36 shows a weir arranged for the purpose of gauging experimentally the quantity of water passing down the stream. A is a plate of thin iron with a notch cut out of it wide enough by estimation to carry off the water with a moderate depth of overfall; this is screwed to a thick plank B, to obtain the requisite stiffness for the plate, and the whole is fixed in the stream as shown. C is a stake with a flat and level top, which is driven into the bed of the stream to such a depth that its top is exactly level with the lip of the weir, and the depth of water flowing over is measured by a common rule held on its summit. The proper distance of the stake from the weir depends on the quantity of water to be dealt with; in small weirs it may be from 1 to 2 feet, in very large ones 20 to 25 feet; the object is to place it far enough away to avoid the curvature of surface which the water suffers as it approaches the weir, as shown by the Figure. There is some uncertainty in measuring by a rule in the manner indicated, arising from the capillary attraction causing the water to adhere to the rule and to rise above its true height. A more correct method is to use Francis's hook-gauge, a rough modification of which is shown by Fig. 36. The stake J is, in this case, driven to such a depth that its top is at a height convenient to the eye, say 30 inches above the level of the lip of the weir; then a rough hook-gauge D, formed of wood about 1 inch thick, is cut in the form shown, the end E being flat and level, and the length E F made exactly equal to G H or 30 inches. The hook-gauge is held against the stake, and carefully adjusted, by the hook at E being first immersed, and then raised until it just coincides with the surface of the water; the depth of overflow is then given by the distance from the top of the stake to the top of the gauge at F, measured by a rule, &c.

(72.) With a thin plate, and depths thus measured from still water, we have the following rules:—

$$G = d \times \sqrt{d} \times l \times 2 \cdot 67$$
$$l = \frac{G}{d \times \sqrt{d} \times 2 \cdot 67}$$
$$d = \left(\sqrt[3]{\frac{G}{l \times 2 \cdot 67}}\right)^2$$

In which G = gallons discharged per minute.
„ d = depth of overflow in inches.
„ l = length of weir in inches.

Thus, with 2 inches overflow, a weir 72 inches long discharges $2 \times 1 \cdot 4142 \times 72 \times 2 \cdot 67 = 543 \cdot 7$ gallons per minute; again, to discharge 694 gallons per minute, with 3 inches overflow, we should require a length of $\frac{694}{3 \times 1 \cdot 732 \times 2 \cdot 67} = 50$ inches; and again, to find the depth of overflow to carry 1282 gallons, with a length of 60 inches, we have $\frac{1282}{60 \times 2 \cdot 67} = 8$, then $\sqrt[3]{8} = 2$, and $2^2 = 4$ inches, the depth required. Table 19 has been calculated by these rules, and its use may be illustrated by the examples just given; thus with 2 inches overflow the Table gives 7·552 gallons per inch, and a weir 72 inches wide will discharge $7 \cdot 552 \times 72 = 543 \cdot 7$ gallons; a weir with 3 inches overflow discharges 13·87 gallons per inch of width, and for 694 gallons we require a length of $\frac{694}{13 \cdot 87} = 50$ inches; a weir 60 inches

WEIRS—TABLE OF DISCHARGE.

long discharging 1282 gallons is equal to $\frac{1282}{60} = 21\cdot36$ gallons per inch wide, which by the Table is due to 4 inches overflow, &c.

TABLE 19.—Of the DISCHARGE of WATER over WEIRS, 1 inch wide, in GALLONS per MINUTE.

Depth.	Gallons.	Depth.	Gallons.	Depth.	Gallons.	Depth.	Gallons.	Depth.	Gallons.
inch.		inch.		inch.		inch.		inch.	
$\frac{1}{8}$	·3338	5	29·85	16$\frac{1}{2}$	179·0	52	1001	89	2242
$\frac{1}{4}$	·6132	5$\frac{1}{8}$	30·97	17	187·1	53	1030	90	2280
$\frac{3}{8}$	·944	5$\frac{1}{4}$	32·12	17$\frac{1}{2}$	195·5	54	1060	91	2318
$\frac{1}{2}$	1·329	5$\frac{3}{8}$	33·26	18	203·9	55	1089	92	2356
$\frac{5}{8}$	1·734	5$\frac{1}{2}$	34·44	19	221·1	56	1119	93	2395
$\frac{3}{4}$	2·185	5$\frac{5}{8}$	35·62	20	238·8	57	1149	94	2433
1	2·670	5$\frac{3}{4}$	36·85	21	256·9	58	1179	95	2472
1$\frac{1}{8}$	3·185	5$\frac{7}{8}$	38·02	22	275·5	59	1210	96	2512
1$\frac{1}{4}$	3·818	6	39·24	23	294·4	60	1241	97	2551
1$\frac{3}{8}$	4·305	6$\frac{1}{8}$	41·72	24	313·9	61	1272	98	2590
1$\frac{1}{2}$	4·905	6$\frac{1}{2}$	44·25	25	333·8	62	1304	99	2630
1$\frac{5}{8}$	5·531	6$\frac{3}{4}$	46·82	26	354·0	63	1335	100	2670
1$\frac{3}{4}$	6·167	7	49·45	27	374·6	64	1367	101	2711
1$\frac{7}{8}$	6·855	7$\frac{1}{4}$	52·12	28	395·6	65	1399	102	2751
2	7·552	7$\frac{1}{2}$	54·84	29	417·0	66	1432	103	2791
2$\frac{1}{8}$	8·271	7$\frac{3}{4}$	57·61	30	438·7	67	1464	104	2825
2$\frac{1}{4}$	9·011	8	60·41	31	460·8	68	1497	105	2873
2$\frac{3}{8}$	9·773	8$\frac{1}{4}$	62·54	32	483·3	69	1531	106	2914
2$\frac{1}{2}$	10·55	8$\frac{1}{2}$	66·17	33	506·1	70	1564	107	2955
2$\frac{5}{8}$	11·36	8$\frac{3}{4}$	69·11	34	529·3	71	1597	108	2997
2$\frac{3}{4}$	12·18	9	72·09	35	552·8	72	1631	109	3039
2$\frac{7}{8}$	13·02	9$\frac{1}{4}$	75·12	36	576·7	73	1665	110	3080
3	13·87	9$\frac{1}{2}$	78·18	37	600·9	74	1700	111	3122
3$\frac{1}{8}$	14·75	9$\frac{3}{4}$	81·29	38	625·4	75	1734	112	3165
3$\frac{1}{4}$	15·64	10	84·43	39	650·4	76	1769	113	3207
3$\frac{3}{8}$	16·55	10$\frac{1}{2}$	90·84	40	675·5	77	1804	114	3250
3$\frac{1}{2}$	17·48	11	97·41	41	700·9	78	1839	115	3293
3$\frac{5}{8}$	18·42	11$\frac{1}{2}$	104·1	42	726·7	79	1875	116	3336
3$\frac{3}{4}$	19·39	12	111·0	43	752·9	80	1910	117	3379
3$\frac{7}{8}$	20·37	12$\frac{1}{2}$	118·0	44	779·3	81	1946	118	3422
4	21·36	13	125·1	45	806·0	82	1983	119	3466
4$\frac{1}{8}$	22·37	13$\frac{1}{2}$	132·5	46	832·8	83	2019	120	3510
4$\frac{1}{4}$	23·39	14	139·8	47	860·3	84	2056	121	3553
4$\frac{3}{8}$	24·38	14$\frac{1}{2}$	147·4	48	887·9	85	2093	122	3598
4$\frac{1}{2}$	25·49	15	155·1	49	915·8	86	2130	123	3642
4$\frac{5}{8}$	26·56	15$\frac{1}{2}$	163·0	50	944·0	87	2162	124	3687
4$\frac{3}{4}$	27·64	16	170·9	51	972·4	88	2204	125	3731
4$\frac{7}{8}$	28·74								

F

(73.) "*Effect of Thickness of Crest.*"—When the lip of the weir has a considerable thickness, which is frequently a practical necessity, the discharge will be less than with a thin plate, a loss arising from friction. Mr. Blackwell's experiments, made on a large scale, and with depths of overfall ranging from 1 inch to 14 inches, give us the following coefficients, by which Table 19 may be adapted to the forms commonly met with in practice:—

	Ratio of Discharge.
Thin plate, weir 10 feet long	1·000
Plank, 2 inches thick, square edged, weirs 3, 6, 10 feet long	·845
Crest, 3 feet thick, level at the top, „ „ „	·712
„ „ sloped top, slope 1 in 12 to 1 in 18	·700

Thus, say we have a river-weir 30 feet wide, with 6½ inches overfall, the crest having a slope of 1 in 12, then the discharge will be 44·25 × 360 × ·76 = 12,107 gallons per minute, or

$$\frac{12107}{6\cdot 23} = 1943 \text{ cubic feet.}$$

(74.) Table 19 may be applied to rectangular apertures like Fig. 35, for the discharge in such a case is the *difference* between two weirs, A, B, C, D, and A, E, F, D; say the head to the top of the aperture (A, B) is 16½ inches, and to the bottom (A, E) 22 inches, and that the width (E, F) is 20 inches. Then, by Table 19, 22 inches = 275·5 gallons per inch, and 16½ inches = 179·0 gallons; the difference is, therefore, 275·5 − 179·0 = 96·5, and the discharge 96·5 × 20 = 1930 gallons; but as *contraction* occurs on four sides in this case, see (51), the real discharge would be 1930 × ·635 ÷ ·667 = 1837 gallons per minute. The coefficients in (73) do not apply to apertures with large heads.

Similarly we may determine the discharge of round apertures, or approximately of any regular figures, which will not differ materially from that of a circumscribing rectangular opening, reduction being made for the true area of the figure whose discharge is required. Thus, say we require the discharge of a circular aperture 12 inches diameter, the head measured from the upper edge of the orifice being 14 inches, therefore, 26 inches above the lower edge. Here we have $354 \cdot 0 - 139 \cdot 8 = 214 \cdot 2$ gallons per inch wide, and if the aperture were rectangular it would discharge $214 \cdot 2 \times 12 = 2570 \cdot 4$ **gallons**; but being circular its area is $\cdot 7854$, that of a circumscribing rectangle being $1 \cdot 0$, and the true discharge is $2570 \cdot 4 \times \cdot 7854 \times \cdot 635 \div \cdot 667 = 1922$ gallons per minute.

(75.) "*Effect of Velocity of Approach to Weirs, &c.*"—We have so far supposed that the head has been measured from still water, or that the channel was of very large area in proportion to the discharging orifices. When the channel is of small area, the water will have a sensible velocity as it approaches the aperture, which will increase the discharge, and correction must be made for it by adding to the measured head, that due to the observed velocity of approach. Table 15 gives the head due to a range of velocities such as are likely to be met with in ordinary practice; thus, in the case of a weir 60 inches wide, with $3\frac{3}{8}$ inches overfall, the discharge $= 18 \cdot 42 \times 60 = 1105 \cdot 2$ gallons, but if the velocity of approach had been 66 feet per minute or $1 \cdot 1$ foot per second, we find the head due to that velocity in Col. B $= \frac{1}{4}$ inch, and the head on the weir becomes $3\frac{3}{8} + \frac{1}{4} = 3\frac{7}{8}$, and the discharge $20 \cdot 37 \times 60 = 1222$ gallons. More strictly, it is the *difference* between two weirs with the respective overfalls of $\frac{1}{4}$ inch and $3\frac{7}{8}$, or $(20 \cdot 37 - \cdot 3338) \times 60 = 1202$ gallons, instead of $1105 \cdot 2$ gallons, as we found it for still water.

(76.) "*Correction for Short Weirs.*"—The rules in (72) assume that the discharge of a weir is simply proportional to its length. This is not strictly correct; in ordinary cases where the weir is narrower than the channel, the issuing stream suffers *contraction* at the two ends, by which its length is virtually reduced, and as this contraction is about the same with all lengths its effect is proportionally greater with short weirs than with long ones. The experiments of Francis show that the effect of contraction at both ends is to reduce the effective length $0 \cdot 2$ inch for each inch in depth of overfall, or 1 inch with 5 inches deep, 2 inches with 10 inches deep, &c. With 5 inches overfall, and weirs

TABLE 20.—The Discharge of Overflow Pipes for Tanks, &c.

Diameter of the Trumpet-mouth in Inches.

Depth of Over-flow in Inches.	2	3	4	5	6	7	8	9	10	11	12	14	16	18
						Gallons Discharged per Minute.								
½	6	9	12	15	18	21	24	27	30	33	36	40	47	53
¾	11	16	22	27	32	38	43	48	54	59	65	75	86	97
1	17	25	34	42	50	59	67	75	84	92	100	118	134	151
1¼	23	35	47	57	70	82	94	106	117	129	140	164	188	211
1½	31	46	62	77	92	108	123	139	154	170	185	216	247	278
1¾	39	58	78	97	116	136	155	175	194	214	233	272	311	350
2	47	71	95	119	142	166	190	214	237	261	285	332	380	427
2¼	..	85	113	142	170	198	227	255	283	312	340	397	453	510
2½	..	100	133	166	199	232	265	299	332	365	398	465	531	597
2¾	..	116	155	194	233	271	310	349	388	427	465	543	621	698
3	..	131	174	218	262	305	349	392	436	480	523	611	698	785
3¼	..	..	197	246	295	344	394	443	492	541	590	689	787	886
3½	..	..	220	275	330	385	440	495	550	605	660	770	880	990
3¾	..	..	244	305	366	427	488	549	610	671	732	854	897	1098
4	..	..	..	336	403	470	537	605	672	740	806	941	1075	1210

5, 10, 20, 50, and 100 inches long, Table 19 gives 149, 298, 597, 1492, and 2985 gallons per minute; but deducting one inch from all those lengths, they are reduced to 4, 9, 19, 49, and 99 inches, and the discharges become 119, 268, 567, 1462, and 2955 gallons. Francis gives a rule for weirs with thin plates, of which the following is a modification:—

$$G = 2\cdot 4953 \times (l - 0\cdot 1\,n\,d) \times d^{\frac{3}{2}}$$

In which n = the number of end contractions (usually two), and the rest as in (72). Where the weir is the full width of the channel, $n = 0$. By this rule, with the real lengths given above, the discharges come out 112, 251, 530, 1367, and 2762 gallons, which are rather less than with the *reduced* lengths by Table 19.

(77.) "*Overflow-pipes to Tanks, &c.*"—The rules and Table for weirs apply also with approximate correctness to an overflow-pipe to a tank, as in Fig. 46, which may be considered as a circular weir whose length is equal to the circumference of the trumpet-mouth. The following rules will give the same result more directly:—

$$G = D \times \sqrt{D} \times d \times 8\cdot 4$$

$$d = \frac{G}{8\cdot 4 \times D \times \sqrt{D}}$$

$$D = \left(\sqrt[3]{\frac{G}{8\cdot 4 \times d}}\right)^2;$$

In which d = the diameter of the trumpet-mouth in inches, D = depth of water over the lip (measured from still-water) in inches, and G = gallons discharged per minute: Table 20 has been calculated by this rule. The size of the discharge-pipe A must be determined by the ordinary rules; with short pipes the discharge is governed principally by the head due to velocity, which is given by Table 1 rather than Table 2 for a pipe of this form. For tanks 3 feet deep, and with a discharge-pipe of that length, Table 21 gives the maximum discharge. Say we had to provide for 400 gallons per minute:—Table 21 shows that

4 inches is the smallest size of pipe admissible, and allowing 2½ inches for overflow, Table 20 gives 12 inches for the least diameter of trumpet-mouth. We must allow some margin for contingencies, and in such a case, the lip of the trumpet-mouth should not be less than 3 inches below the top of the tank, and thus 3 inches is practically lost in the useful depth of the tank.

TABLE 21.—Of the MAXIMUM DISCHARGE of VERTICAL PIPES 3 FEET LONG.

Diameter of Pipe in Inches.	Maximum Discharge in Gallons per Minute.	Diameter of Pipe in Inches.	Maximum Discharge in Gallons per Minute.
1	19	3½	303
1½	45	4	400
2	88	5	630
2½	145	6	920
3	220	7	1300

(78.) Fig. 47 shows a simple contrivance of the late Mr. Appold, by which this loss may be avoided, and the water-level not allowed to rise more than about ⅛th of an inch above the lip of the trumpet-mouth, even when the descending pipe is discharging full-bore. B is a dished cover of sheet copper, &c., supported on four brackets C, C, cast on the pipe, so that its lip at D is at the same level as the lip of the trumpet-mouth. When the water rises to that level, it does not immediately flow over when the lip is dry, but rises perhaps $\frac{1}{10}$th of an inch above it, and then, suddenly overflowing, creates a partial vacuum under the cover, causing the water to rise there above the level of the water in the tank and filling the pipe full-bore. The air under the cover is swallowed up by the rush of the water, and the maximum quantity which the pipe can carry is delivered. This continues till, the water being drawn down below the lip of the cover at D, air enters, and the action suddenly ceases, to be again repeated should the water rise. As the action depends on the suction of the down-pipe, which will not be perfect if the bore is not well filled, it is expedient not to make that pipe *much* larger than necessary. It is usual to construct the pipe so as to serve as a wash-out valve, the joint at the bottom being turned and bored to fit water-tight.

(79.) "*Overflows to Fountains.*"—In ornamental fountains with shallow basins it is important that the water-level should fluctuate as little as possible; hence the form of overflow-pipe just described is specially applicable to such cases. It is generally desirable that the pipe should be concealed, which may be done by fixing it in a small supplementary cistern by the side of the fountain basin, with a large passage between them. For small fountains with say 100 gallons per minute, an inverted overflow-pipe may be used, as in Fig. 42; a short pipe A, which serves also as a waste-pipe to empty the basin when necessary by the cock B, carries the overflow trumpet-mouth C. Say we have 100 gallons; then with a 6-inch pipe at A, the head for velocity at entry would be about 1 inch, and with a 12-inch trumpet-mouth the head for overflow, by Table 20, is also 1 inch, so that the water-line would fluctuate 2 inches. The cock B may be of smaller size, say 3 inches, the end of the pipe being reduced to suit it. With care, such an arrangement might be used for a very large quantity, by adjusting the cock so as to carry rather less than the supply, leaving the trumpet-mouth to carry off the surplus and regulate the level.

(80.) "*Common Overflow-pipe.*"—When an overflow takes the form of a short pipe inserted in the side of a cistern, as in Fig. 45, and the water to be carried off is just sufficient to fill the pipe, the discharge will be given approximately by the following rule:—

$$G = d^{2\cdot 5} \times 3\cdot 2;$$

In which G = gallons discharged per minute.
,, d = diameter in inches.

Table 22, which has been calculated by this rule, may also be useful for another purpose. It sometimes happens that the only datum which an engineer obtains as a basis for rough estimates is, that a spring or stream delivers "about as much as a pipe of a certain size would carry." This, of course, is very indefinite, but in most cases it means the amount which a pipe would discharge without **extra** pressure, as in Fig. 45 and Table 22: thus an 8-inch pipe just filled delivers about 580 gallons per minute:—the pipe in (37) was observed to be nearly filled with the issuing stream when discharging 564 gallons.

TABLE 22.—Of the DISCHARGE of OUTLET-PIPES, Fig. 45.

Diameter. Inches.	Gallons per Minute.	Diameter. Inches.	Gallons per Minute.	Diameter. Inches.	Gallons per Minute.
1	3·2	5	179	13	1950
1½	8·8	6	283	14	2346
2	18·1	7	415	15	2788
2½	31·6	8	580	16	3277
3	50·0	9	778	17	3814
3½	73·3	10	1012	18	4400
4	112·4	11	1284	19	5037
4½	138·0	12	1600	20	5725

CHAPTER V.

ON THE STRENGTH OF WATER-PIPES — RAINFALL, &c., &c.

(81.) "*Strength of Thick Pipes.*"—The strength of pipes to resist an internal pressure is not simply proportional to the thickness of metal. The material stretches or extends under a tensile strain, and the result of extension is, that the inside metal is more strained than that of the outside, and that thick pipes are weaker in proportion to their thickness than thin ones. Barlow has given the following rules :—

$$T = \frac{R + P}{S - P}$$

$$P = \frac{S \times T}{R + T}$$

$$S = \frac{(R + T) \times P}{T};$$

In which S = the cohesive strength of the metal per square inch.
,, P = the internal pressure per square inch, in the same terms as S.
,, R = the radius of the inside of the pipe in inches.
,, T = the thickness of metal in inches.

For cast-iron S may be taken at 7·142 tons, or 16,000 lbs. per square inch, and with that strength we obtain the bursting pressure given by Table 23, which shows that with a 10-inch pipe a thickness of 10 inches gives only four times the strength due to a thickness of 1 inch.

TABLE 23.—Of the STRENGTH of a 10-INCH CAST-IRON PIPE to RESIST INTERNAL PRESSURE, in Tons per Square Inch.

Thickness in inches	1	2	3	4	5
Pressure by Barlow's rule	1·19	2·04	2·68	3·17	3·57
Pressure by exact calculation	1·226	2·161	2·896	3·485	3·972
Thickness in inches	6	7	8	9	10
Pressure by Barlow's rule	3·90	4·17	4·40	4·59	4·76
Pressure by exact calculation	4·337	4·722	5·019	5·275	5·5

Barlow's rule supposes that the extensions are simply proportional to the strain, which is not quite correct; by taking the true extensions we obtain the second series of bursting pressures given in the Table by a calculation which need not be here elaborated.

(82.) "*Strength of Thin Pipes.*"—Barlow's rule is quite inapplicable to comparatively thin pipes, such as are commonly used for water and gas; there are other and practical considerations which that rule does not contemplate. With thin pipes and moderate pressures, we have to consider not only the thickness necessary to bear the pressure, but also that required to bear the traffic along the roads in which they are commonly laid. Again, although great care is taken to keep the core central, it is seldom perfectly so; a pipe intended to be ½-inch thick is frequently

THICKNESS AND WEIGHT OF CAST-IRON SOCKET-PIPE.

TABLE 24.—Of the THICKNESS and WEIGHT of CAST-IRON SOCKET-PIPE to BEAR SAFELY DIFFERENT PRESSURES of WATER.

Diameter in Inches.	Length exclusive of Socket.		For Gas, &c.			100 feet.			250 feet.			500 feet.			750 feet.			1000 feet.		
	ft.	in.	thick.	cwt.	qrs. lbs.	thick.	cwt.	qrs. lbs.	thick.	cwt.	qrs. lbs.	thick.	cwt.	qrs. lbs.	thick.	cwt.	qrs. lbs.	thick.	cwt.	qrs. lbs.
1¼	6	0	·27	0	1 3	·28	0	1 4	·29	0	1 5	·30	0	1 7	·31	0	1 8	·33	0	1 10
2	6	0	·29	0	1 17	·30	0	1 19	·31	0	1 20	·33	0	1 23	·35	0	1 26	·37	0	2 2
2½	6	0	·30	0	2 1	·31	0	2 3	·33	0	2 7	·35	0	2 11	·37	0	2 14	·40	0	2 20
3	9	0	·32	0	3 18	·33	0	3 21	·35	1	0 3	·38	1	0 9	·41	1	0 19	·44	1	1 9
4	9	0	·35	0	1 7	·37	1	1 15	·39	1	1 24	·43	1	2 13	·47	1	3 1	·51	1	3 18
5	9	0	·37	1	2 23	·39	1	3 5	·42	1	3 21	·47	2	0 19	·52	2	1 16	·57	2	2 14
6	9	0	·39	2	0 16	·42	2	1 6	·45	2	1 25	·51	2	3 6	·57	3	0 15	·63	3	1 24
7	9	0	·41	2	2 13	·44	2	3 8	·48	3	0 9	·55	3	2 4	·62	4	0 0	·69	4	1 21
8	9	0	·43	3	0 14	·46	3	1 10	·51	3	2 23	·59	4	1 1	·67	4	3 13	·75	5	1 22
9	9	0	·45	3	2 18	·48	3	3 17	·53	4	1 7	·63	5	0 14	·72	5	3 12	·81	6	2 10
10	9	0	·47	4	0 26	·51	4	2 10	·57	5	0 15	·67	6	0 4	·77	6	3 21	·87	7	3 9
12	9	0	·49	5	1 2	·54	5	3 6	·61	6	2 6	·73	7	3 11	·85	9	0 15	·97	10	2 0
15	9	0	·53	7	1 0	·59	8	0 6	·68	9	1 4	·83	11	1 9	·98	13	1 14	1·13	15	2 0
18	9	0	·57	9	1 0	·64	10	1 16	·75	12	1 0	·93	15	0 11	1·11	18	0 0	1·29	21	0 10
21	9	0	·60	11	0 11	·69	12	3 12	·81	15	0 18	1·02	19	1 7	1·23	23	1 0	1·44	27	1 3
24	9	0	·64	13	2 0	·73	15	2 2	·88	18	3 2	1·12	23	2 4	1·36	28	2 9	1·60	33	2 14
30	9	0	·69	18	0 14	·81	21	1 6	1·00	26	1 2	1·29	34	3 4	1·59	42	2 18	1·89	50	2 5
36	9	0	·75	23	2 16	·89	28	0 6	1·11	35	1 0	1·47	46	1 10	1·83	57	3 0	2·19	69	1 12

⅞ths at one side and ⅝ths at the other, and of course the least thickness governs the strength of the pipe. And again, there are in most cases shocks arising from the closing of cocks, &c., against which it is necessary to provide adequate strength. In thin pipes, therefore, the determination of the thickness becomes a practical question, and we must obtain an empirical rule from experience. The rule may take the following form:—

$$t = \left(\frac{\sqrt{D}}{10} + \cdot 15\right) + \left(\frac{H \times D}{2500}\right);$$

In which D = the diameter of the pipe in inches.
„ H = the safe head of water, in feet.
„ t = the thickness of metal in inches.

Table 24 has been calculated by this rule, and we have also given the approximate weights, from gas-pipes in which the pressure is practically nothing, up to 1000 feet of water. Engineers usually specify the *weight* of their pipes rather than the thickness, leaving the founder to fix that for himself, which long practice enables him to do with considerable precision. Of course absolute correctness cannot be attained, and should not be expected; a margin should be allowed, say one pound to the inch, either way; so that, for instance, a 10-inch pipe for 100 feet head, specified to weigh 4 cwt. 2 qrs. 10 lbs., as per Table 24, should not be rejected if its real weight is between 4 cwt. 2 qrs. 0 lbs. and 4 cwt. 2 qrs. 20 lbs., &c. Founders consider this to be a fair allowance for variation in weight.

(83.) "*Proportions of Socket-pipes.*"—The joints of water-pipes are usually made by sockets and spigots run with melted lead; and this is the best mode. Such pipes are easy to cast, and consequently cheap, the joints are more easily made than with flanges, and they admit a considerable departure from the strictly straight line which is sometimes very convenient. But to allow for this the sockets must be made of larger diameter than is necessary where the line is straight, and for this reason, perhaps, sockets are frequently made larger than they should be for making a good joint. For ordinary cases ¼ inch in thickness or ½ inch in diameter will suffice for pipes of 3 inches diameter and under; say $\frac{5}{16}$ from 3 to 10 inches; and $\frac{3}{8}$ for larger sizes. Table 25 gives the general proportions for socket-joints, weight of

TABLE 25.—Of the PROPORTIONS of JOINTS, &c., for CAST-IRON SOCKET-PIPES.

Diameter of Pipe in Inches.	Depth of Socket.	Lead-joint.			Laying per yard. Prime Cost.
		Thickness.	Depth.	Weight in lbs.	
	inches.				s. d.
$1\frac{1}{2}$	3	$\frac{1}{4}$	$1\frac{1}{2}$	1·2	0 11
2	3	$\frac{1}{4}$	$1\frac{1}{2}$	1·4	1 0
$2\frac{1}{2}$	$3\frac{1}{4}$	$\frac{1}{4}$	$1\frac{1}{2}$	1·6	1 1
3	$3\frac{1}{2}$	$\frac{1}{4}$	$1\frac{3}{4}$	2·3	1 2
4	4	$\frac{5}{16}$	2	4·0	1 3
5	4	$\frac{5}{16}$	2	5·0	1 5
6	$4\frac{1}{4}$	$\frac{5}{16}$	$2\frac{1}{4}$	6·5	1 7
7	$4\frac{1}{4}$	$\frac{5}{16}$	$2\frac{1}{4}$	7·7	1 10
8	$4\frac{1}{2}$	$\frac{5}{16}$	$2\frac{1}{4}$	8·2	2 1
9	$4\frac{1}{2}$	$\frac{5}{16}$	$2\frac{1}{2}$	10·4	2 6
10	$4\frac{1}{2}$	$\frac{5}{16}$	$2\frac{1}{2}$	11·5	3 4
12	$4\frac{1}{2}$	$\frac{3}{8}$	$2\frac{3}{4}$	18·0	4 6

lead, &c.: we have also added the average cost of laying pipes, including excavating the ground and making good the same; this will vary of course with the nature of the ground and the cost of labour in different localities.

In Table 26 we have given the weights of socket-pipes and connections by Bailey, Pegg, and Co., of Bankside, Southwark: by reference to Table 24 it will be seen that these pipes are of a weight and strength suitable for about 150 feet head in the larger sizes, and 250 feet in the smaller ones.

(84.) "*Proportions of Flange-pipes.*"—Flange-pipes are not very often used for water, for reasons already given; but they are convenient for temporary purposes, where the joints have to be frequently broken. Table 27 gives the best proportions for the flanges, bolts, &c., which will be found to differ considerably from those adopted by many makers. The flanges of cast-iron pipes are frequently made excessively large in diameter and very light in metal. India-rubber rings form the most convenient kind of joint for flange-pipes.

TABLE 26.—Of the **WEIGHT, &c., of Ordinary** (Stock) SOCKET-PIPES, **BENDS,** CONNECTIONS, **&c.**

Diameter of Pipe.	Diameter of Socket.	Length without Socket.	Pipe.	Quarter Bends 90°.	Eighth Bends 45°.	Branches. Fig. 48.	Outlets or Tees. Fig. 49.	Double Collars.	Caps.
inches.	inches.	ft. in.	cwt. qrs. lbs.	cwt. qrs. lbs.	cwt. qrs. lbs.	cwt. qrs. lbs.	cwt. qrs. lbs.	cwt. qrs. lbs.	cwt. qrs. lbs.
1½	2⅝	6 0	0 1 8	0 0 15	0 0 13	0 1 12	0 1 5	0 0 11	0 0 6
2	3⅛	6 0	0 2 0	0 1 1	0 0 23	0 1 26	0 1 15	0 0 16	0 0 8
2¼	3½	6 0	0 2 7	0 1 6	0 0 25	0 2 24	0 2 0	0 0 24	0 0 9
3	4⅛	9 0	1 0 0	0 1 11	0 1 6	0 3 2	0 2 24	0 0 26	0 0 13
4	5¼	9 0	1 2 0	0 2 5	0 1 25	1 0 16	0 3 21	0 1 10	0 0 23
5	6¼	9 0	2 0 0	1 0 6	0 3 16	1 1 21	1 0 24	0 1 12	0 1 0
6	7½	9 0	2 2 0	1 1 24	1 1 6	1 2 21	1 2 1	0 2 0	0 1 7
7	8½	9 0	3 0 6	1 3 0	1 3 4	2 1 0	2 0 21	0 2 22	0 1 21
8	9⅝	9 0	3 2 7	2 3 0	2 0 0	2 3 0	2 2 14	0 3 20	0 2 8
9	10¾	9 0	4 0 7	3 0 7	2 2 0	3 1 0	3 2 0	0 3 21	0 2 21
10	11⅞	9 0	4 3 0	3 1 14	2 3 0	4 1 20	4 0 7	1 0 3	0 3 11
12	13⅞	9 0	6 0 7	3 3 14	3 1 14	5 2 0	5 1 0	1 1 0	1 1 21

TABLE 27.—Of the Proportions of Cast-iron Flange-pipes.

Diameter of Pipe.	Diameter of Flange.	Thickness of Flange.	No. of Bolts.	Diameter of Bolts.	Diameter of Circle of Bolts.
inches.	inches.	inches.		inches.	inches.
$1\frac{1}{2}$	$4\frac{1}{2}$	$\frac{1}{2}$	3	$\frac{3}{8}$	$3\frac{1}{4}$
2	$5\frac{1}{4}$	$\frac{1}{2}$	3	$\frac{7}{16}$	$3\frac{3}{4}$
$2\frac{1}{2}$	6	$\frac{5}{8}$	4	$\frac{7}{16}$	$4\frac{1}{2}$
3	$6\frac{1}{2}$	$\frac{5}{8}$	4	$\frac{1}{2}$	5
4	8	$\frac{5}{8}$	4	$\frac{9}{16}$	$6\frac{1}{4}$
5	$9\frac{1}{4}$	$\frac{3}{4}$	4	$\frac{9}{16}$	$7\frac{1}{2}$
6	$10\frac{1}{2}$	$\frac{3}{4}$	6	$\frac{9}{16}$	$8\frac{3}{4}$
7	12	$\frac{3}{4}$	6	$\frac{5}{8}$	10
8	$13\frac{1}{2}$	$\frac{7}{8}$	6	$\frac{5}{8}$	$11\frac{1}{2}$
9	$14\frac{1}{2}$	$\frac{7}{8}$	6	$\frac{5}{8}$	$12\frac{1}{4}$
10	16	1	6	$\frac{3}{4}$	$13\frac{1}{2}$
12	$18\frac{1}{4}$	1	6	$\frac{3}{4}$	16

(85.) "*Strength of Lead Pipes.*"—The strength of lead pipe may be calculated by Barlow's rule (81), taking the cohesive strength of drawn lead at 2745 lbs. per square inch, as determined by direct experiment. Lead pipes are made of various weights to suit the varying requirements of practice; taking medium weights, and deducing the thickness therefrom, we obtain the following Table, in which the safe working pressure is taken at $\frac{1}{15}$th of the bursting strain :—

Diameter of pipe	$\frac{1}{2}$	$\frac{5}{8}$	$\frac{3}{4}$	1	$1\frac{1}{4}$	$1\frac{1}{2}$	$1\frac{3}{4}$	2
Weight of pipe, lbs. per foot	1·33	1·47	1·87	2·80	4·33	6·0	6·75	8·0
Safe pressure, feet of water	232	183	174	151	152	140	122	116

(86.) "*Power of Horses, &c., in raising Water.*"—The power of men, horses, &c., in raising water varies with the duration of the labour. The following Table gives the number of gallons raised 1 foot high per minute, with common deep-well pumps, and the mean velocity in feet per minute.

Velocity.	Hours per Day.	4.	5.	6.	8.	10.
176	Horse, walking in a circle	1653	1480	1350	1169	1040
180	Pony, or mule, ditto	1102	986	898	780	697
120	Bullock, ditto	1470	1314	1200	1040	930
157	Ass, ditto	457	410	374	323	290
220	Man, with winch pump	249	222	203	176	157
147	Ditto, Contractor's pump	205	183	167	145	130

A good high-pressure steam-engine should raise 3300 gallons 1 foot high per minute per nominal horse-power; the friction of the pumps being compensated by the excess of the indicated power over the nominal.

(87). "*Rainfall.*"—The depth of rain in this country varies very much with the locality; the east coast is the driest, the annual rainfall being in Northumberland about 28·67 inches, diminishing thence gradually to 23 in Norfolk and to 19·8 in Essex, which is the minimum. Thence southward and westward it gradually increases to 25·6 in Kent, 30.64 in Sussex, 38·75 in Dorset, 48·3 in Devon, and 50·6 in Cornwall. The midland districts have a medium fall: Middlesex 24·1, Leicester 26·0, Hereford 29·27, Cheshire 31·3, &c., &c.

"*Heavy Rains.*"—For town drainage and other purposes, we require to know the maximum fall of rain during storms. We find that in

1   5   15   30   45   60   120   180 minutes

the maximum fall of rain may be

0·2   0·75   1·0   1·8   2·5   3·25   3·6   4 inches,

which is at the rate per hour of

12   9   4   3·6   3·3   3·25   1·8   1·33 inches.

"*Rain-water Tanks.*"—Where it is desired to utilize as much as possible of the rain falling on a building, the minimum size of tank becomes an important but complicated question. Taking a place with 24 inches annual rainfall, we have evidently an allowance for a regular consumption of 2 inches per month. But there may be a drought in which for one month no rain falls, and the tank must have 2 inches in store to supply the deficiency. There may also be a wet month with 6 inches of rain, and as only 2 inches is consumed, 4 inches must be stored. The tank must therefore hold $2 + 4 = 6$ inches, or $\frac{1}{4}$th of the annual rainfall. Again, for two months we require 4 inches, but the rainfall varies from $1\frac{1}{2}$ to $7\frac{1}{2}$ inches, and the tank must hold $(4 - 1\frac{1}{2}) + (7\frac{1}{2} - 4) = 6$ inches, as before. For three months we require 6 inches, but the rainfall varying from 2·4 to 8·7 inches, the tank should hold $(6 - 2·4) + (8·7 - 6) =$

6·3 inches. From all this we find that a rain-water tank should hold at least ¼th of the annual rainfall. Thus, with 24 inches, or 2 feet per year, a building 1830 square feet in area, collects 1830 × 2 = 3660 cubic feet, allowing a consumption of 10 cubic feet, or 62·3 gallons per day, and the tank should hold 3660 ÷ 4 = 915 cubic feet.

(88.) "*Weight and Pressure of Water.*"—A gallon of water at 62° weighs 10 lbs., and contains 277·274 cubic inches, or ·16046 cubic foot: hence a cubic foot weighs 62·321 lbs., and contains 6·2321, or nearly 6¼ gallons. Table 28 gives the **pressure in** pounds per **square** inch due to given columns of water and mercury.

TABLE 28.—Of EQUIVALENT PRESSURES in POUNDS per SQUARE INCH, FEET of WATER, and INCHES of MERCURY, at a Temperature of 62° Fahr.

Pounds per Square Inch.	Feet of Water.	Inches of Mercury.	Pounds per Square Inch.	Feet of Water.	Inches of Mercury.
1·	2·311	2·046	2·5962	6·	5·31198
2·	4·622	4·092	3·0289	7·	6·19731
3·	6·933	6·138	3·4616	8·	7·08264
4·	9·244	8·184	3·8942	9·	7·96797
5·	11·555	10·230	·48875	1·12952	1·
6·	13·866	12·276	·97750	2·25904	2·
7·	16·177	14·322	1·46625	3·38856	3·
8·	18·488	16·368	1·95500	4·51808	4·
9·	20·800	18·414	2·44375	5·64760	5·
·4327	1·	·88533	2·93250	6·77712	6·
·8654	2·	1·77066	3·42125	7·90664	7·
1·2981	3·	2·65599	3·91000	9·03616	8·
1·7308	4·	3·54132	4·39875	10·16568	9·
2·1635	5·	4·42665			

EXAMPLE.—Required the **Pressure per Square** Inch, and Equivalent **Column** of Mercury for a Head **of 247 feet of Water**.

Feet of Water.		Pounds per Square Inch.		Inches of Mercury.
200	=	86·54	or	177·066
40	=	17·308	„	35·413
7	=	3·029	„	6·197
247	=	106·877	„	218·676

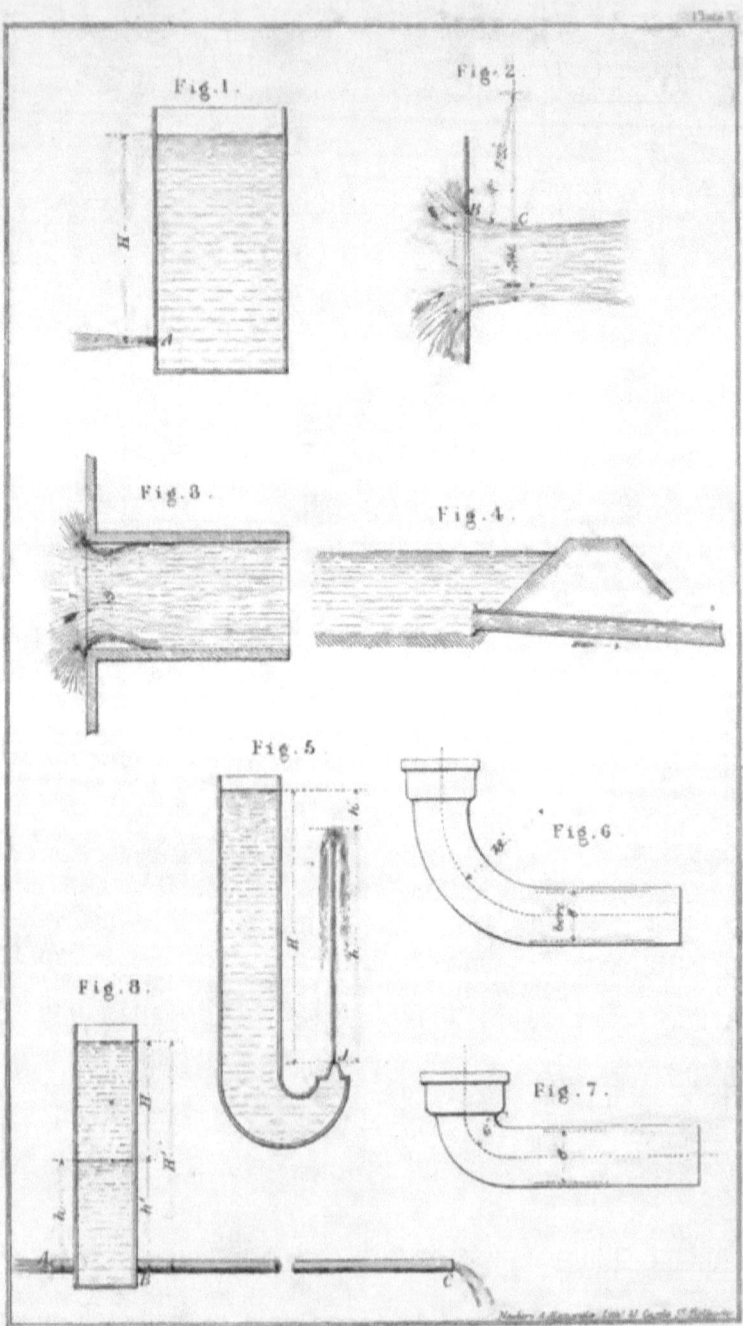

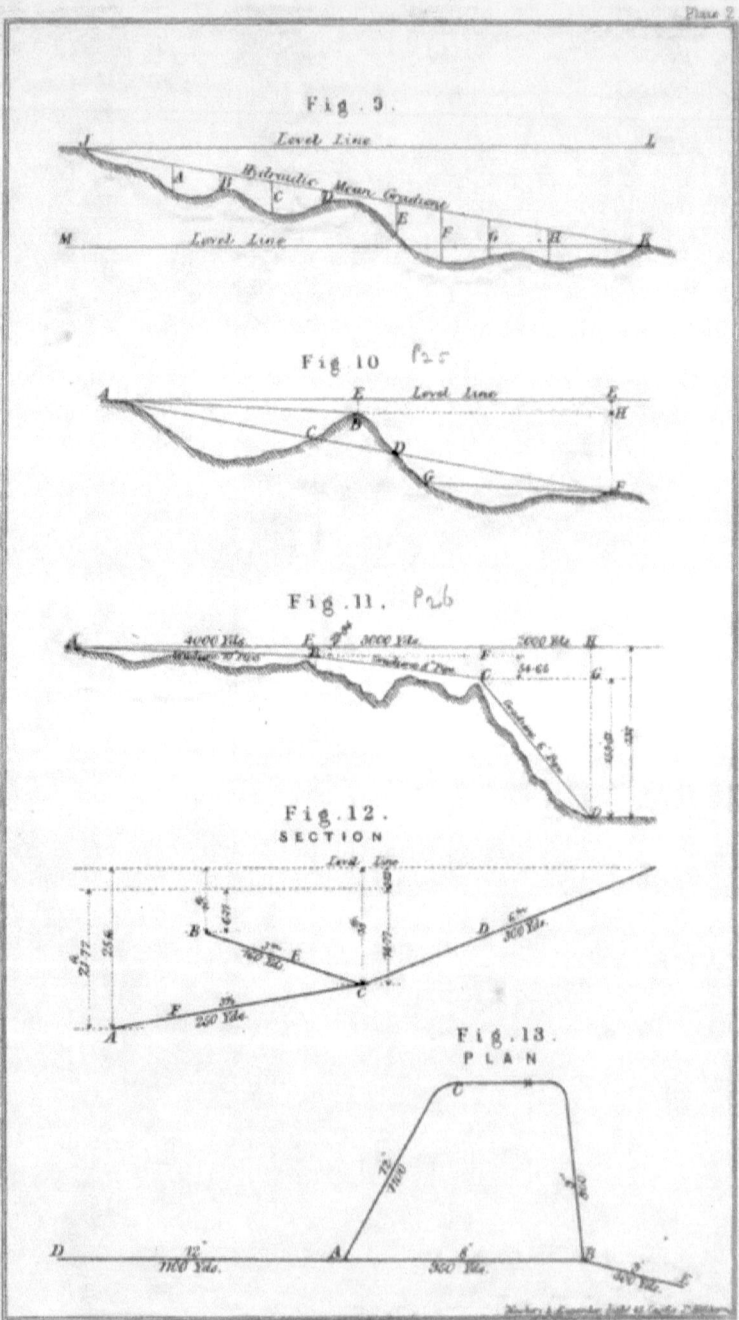

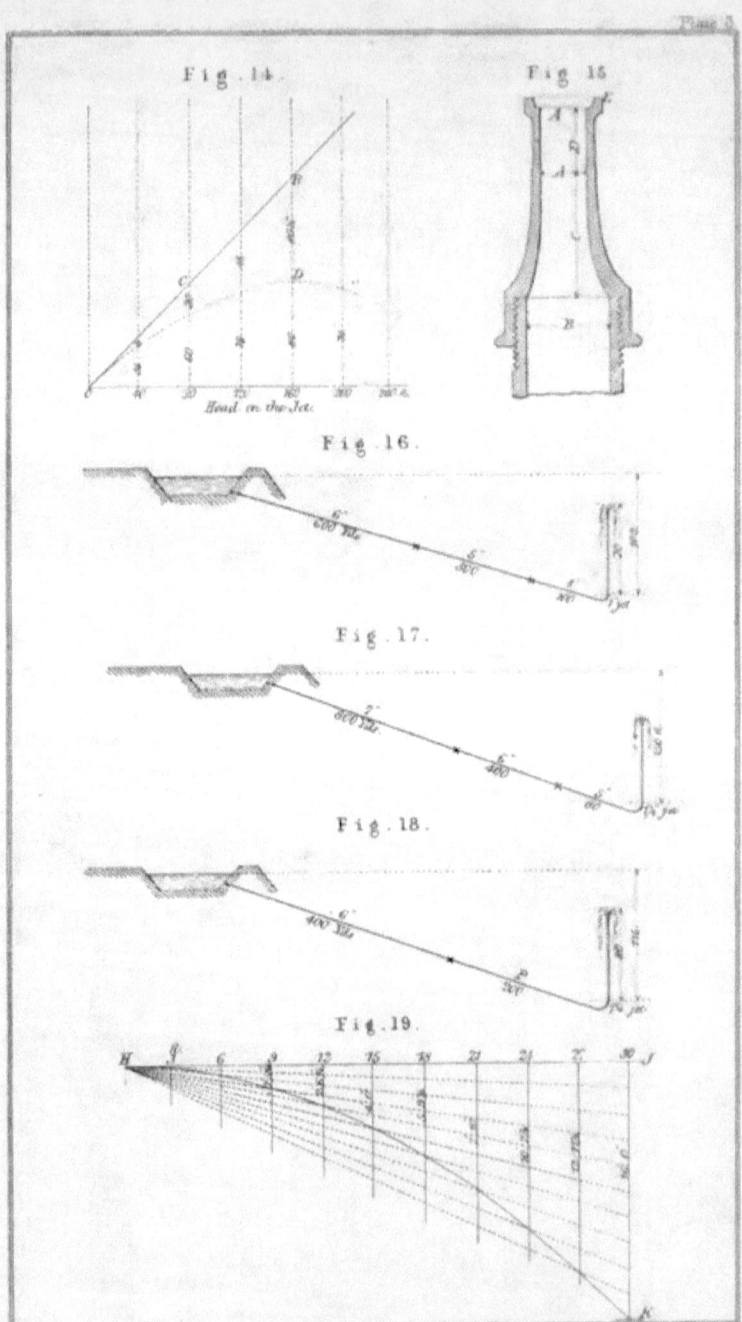

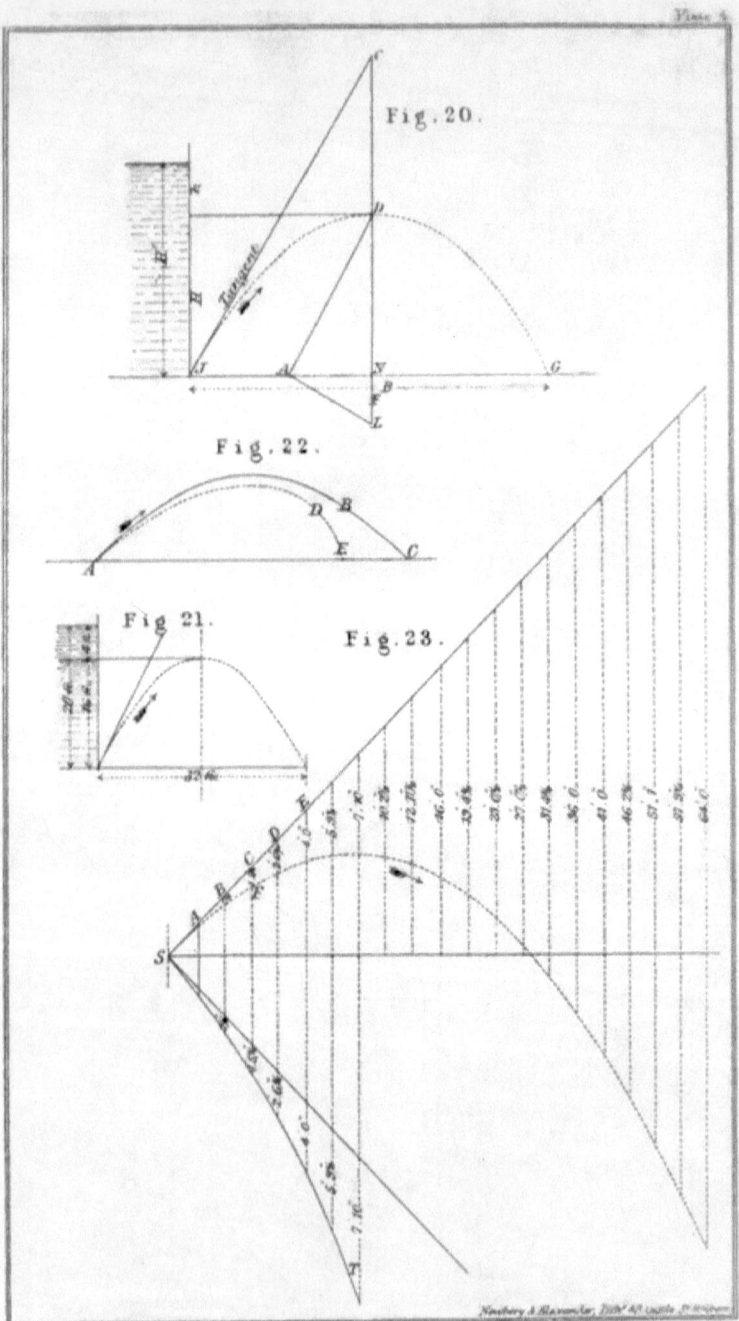

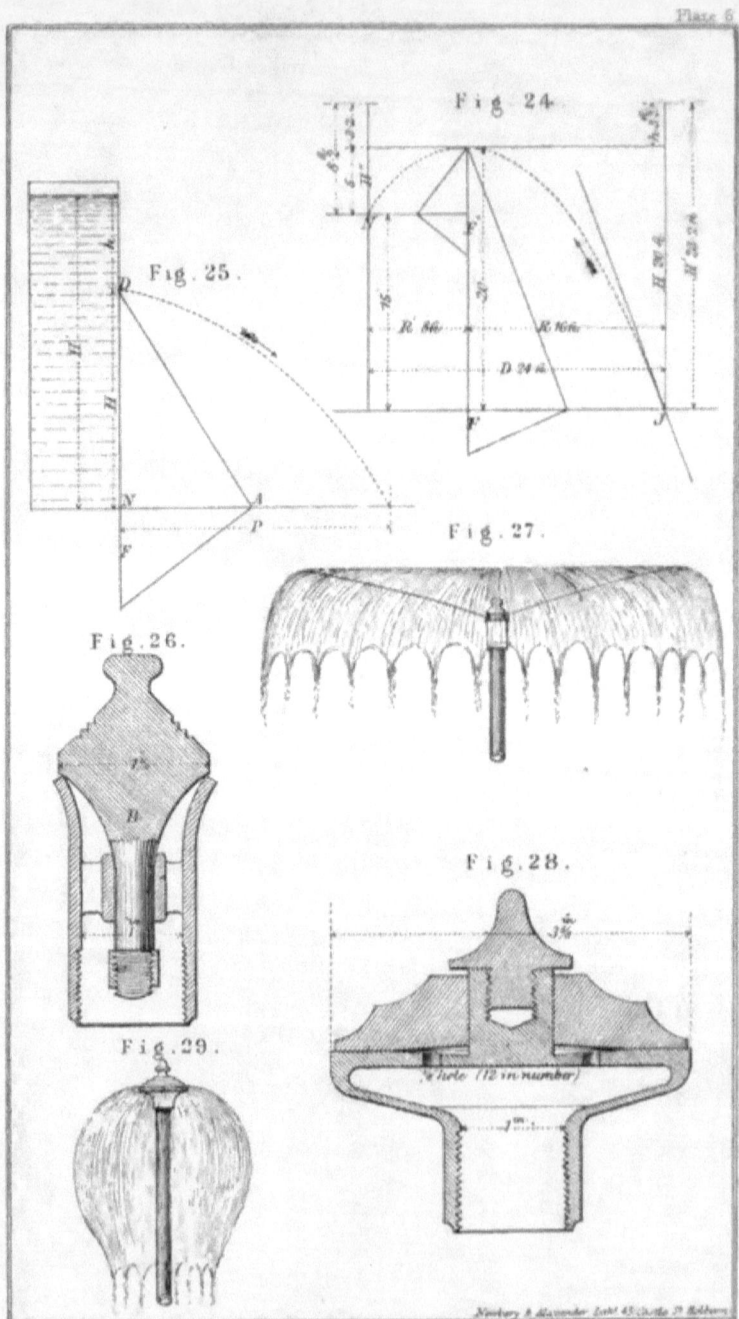

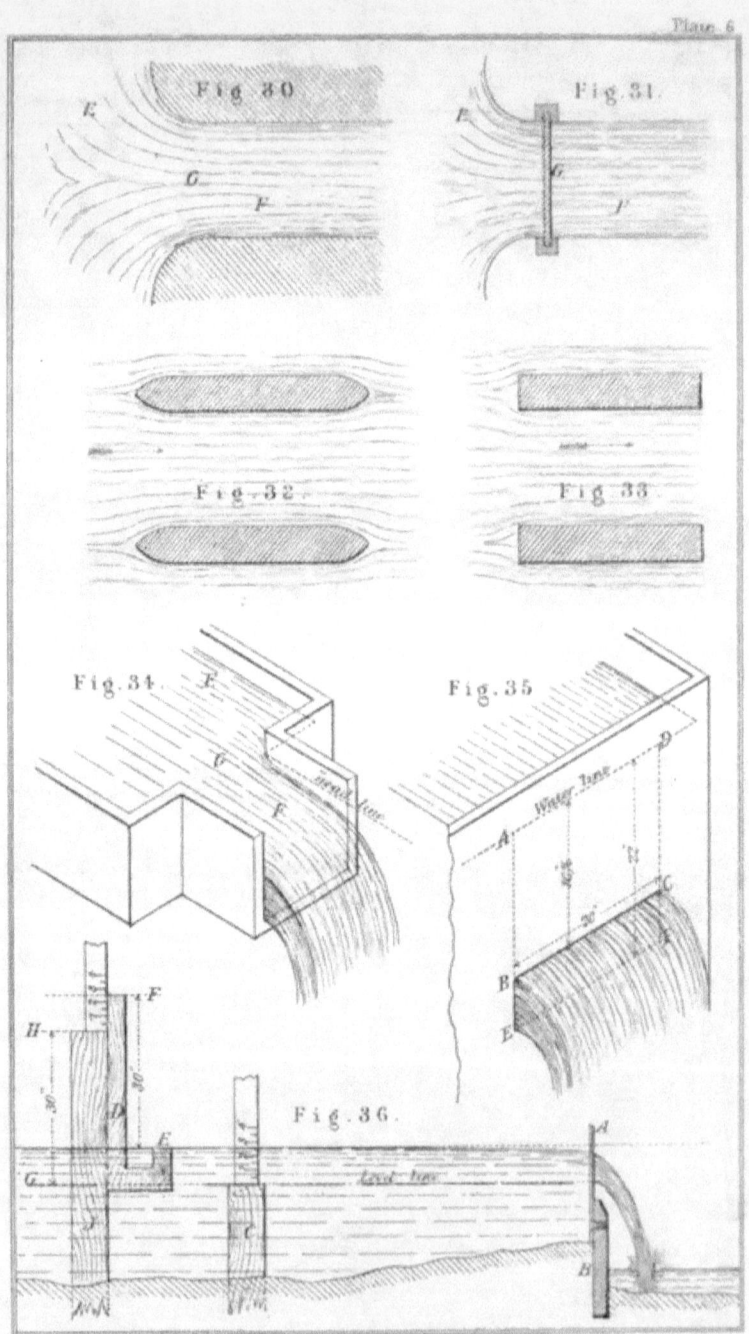

Plate 6.

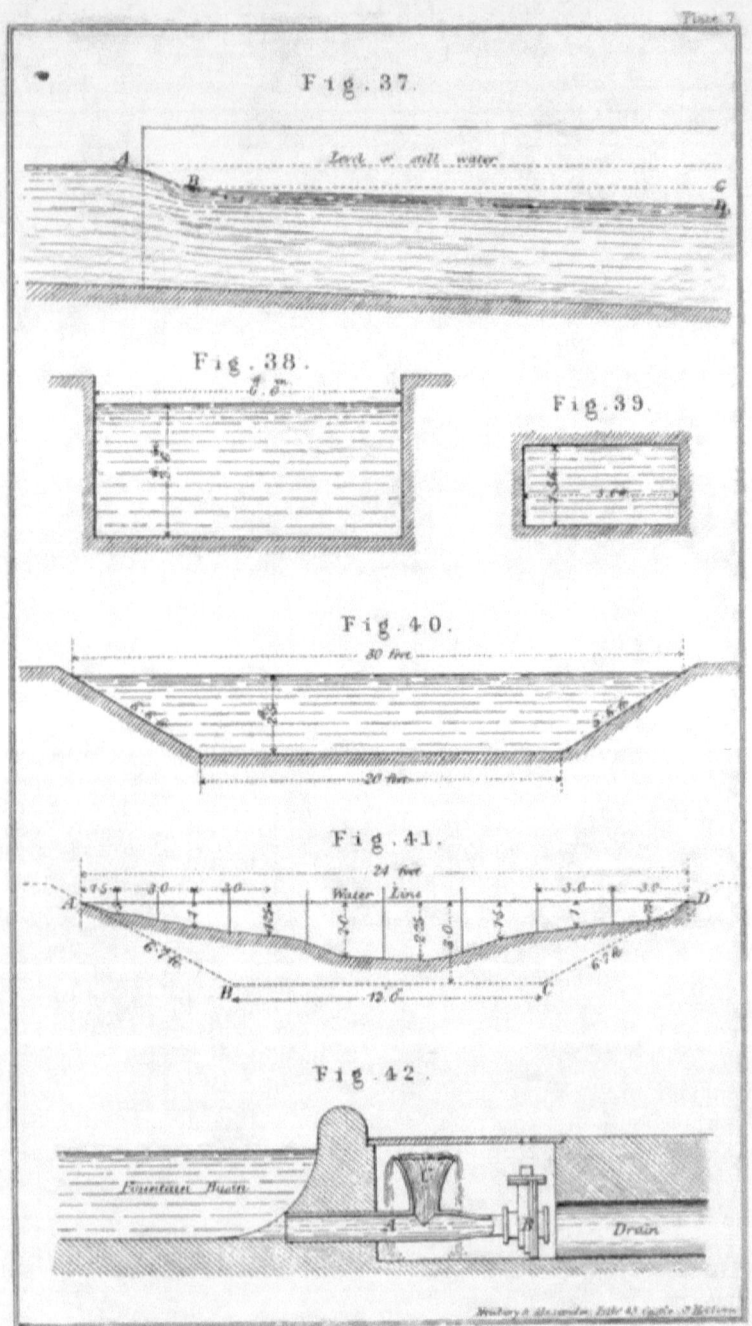

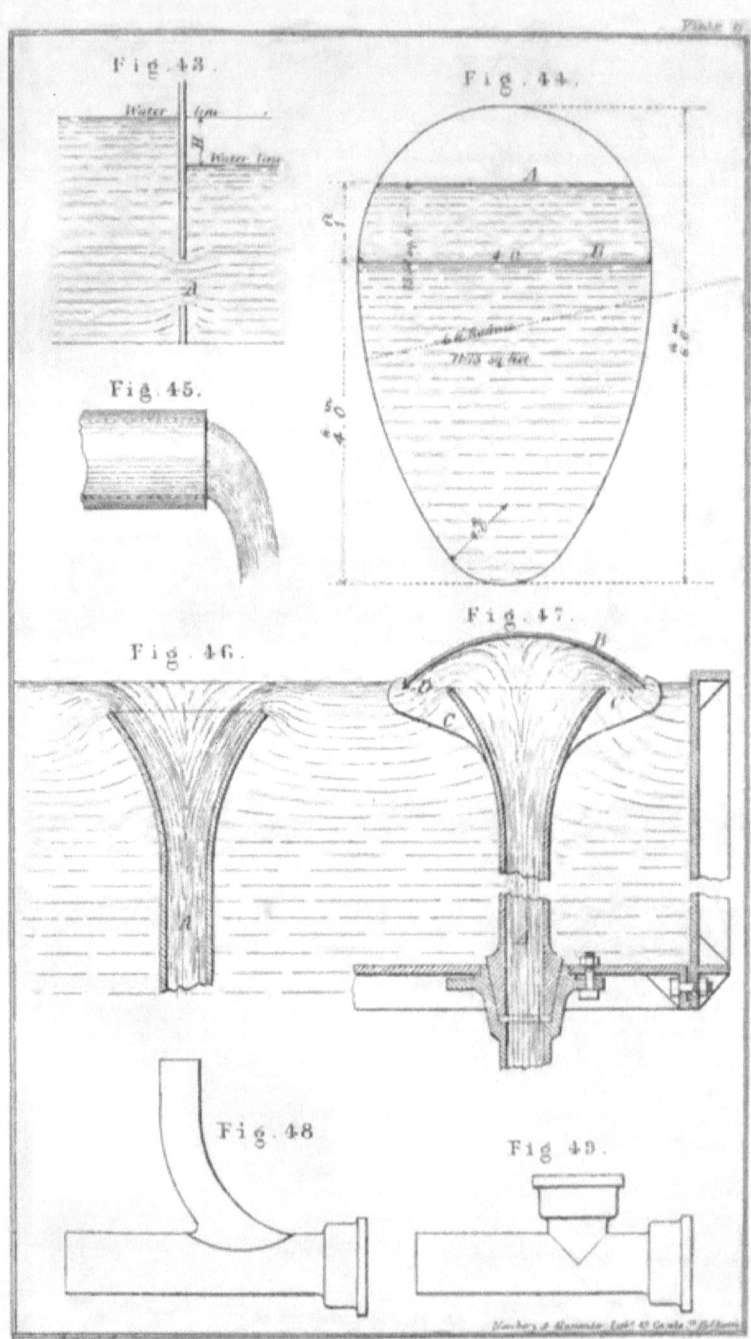

www.ingramcontent.com/pod-product-compliance
Lightning Source LLC
Chambersburg PA
CBHW021946160426
43195CB00011B/1246